Modern Molecular Biology

Applied Bioinformatics and Biostatistics in Cancer Research

Series editor: Jeanne Kowalski, Steven Piantadosi

For other titles published in this series, go to
www.springer.com/series/7616

Srinivasan Yegnasubramanian · William B. Isaacs
Editors

Modern Molecular Biology

Approaches for Unbiased Discovery in Cancer Research

Springer

Editors
Srinivasan Yegnasubramanian
Department of Oncology
Sidney Kimmel Comprehensive
Cancer Center
Johns Hopkins University,
School of Medicine
Baltimore, MD
USA
syegnasu@jhmi.edu

William B. Isaacs
Department of Urology,
Pharmacology, Oncology
Sidney Kimmel Comprehensive
Cancer Center
Brady Urological Institute
Johns Hopkins University,
School of Medicine
Baltimore, MD
USA
wisaacs@jhmi.edu

ISBN 978-0-387-69744-4 e-ISBN 978-0-387-69745-1
DOI 10.1007/978-0-387-69745-1
Springer New York Dordrecht Heidelberg London

Library of Congress Control Number: 2010933871

Printed on acid-free paper

Springer is part of Springer Science+Business Media (www.springer.com)

Preface

Molecular biology has rapidly advanced since the discovery of the basic flow of information in life, from DNA to RNA to proteins. While there are several important and interesting exceptions to this general flow of information, the importance of these biological macromolecules in dictating the phenotypic nature of living creatures in health and disease is paramount. In the last one and a half decades, and particularly after the completion of the Human Genome Project, there has been an explosion of technologies that allow the broad characterization of these macromolecules in physiology, and the perturbations to these macromolecules that occur in diseases such as cancer. In this volume, we will explore the modern approaches used to characterize these macromolecules in an unbiased, systematic way. Such technologies are rapidly advancing our knowledge of the coordinated and complicated changes that occur during carcinogenesis, and are providing vital information that, when correctly interpreted by biostatistical/bioinformatics analyses, can be exploited for the prevention, diagnosis, and treatment of human cancers. The primary purpose of this volume is to help bridge the gap between molecular biologists/cancer researchers and bioinformatics/computational biology researchers by providing an overview of these technologies to those that are not yet familiar with them. With this in mind, we provide an introduction to these technologies and showcase how these have been used to gain an understanding of each of the major macromolecules that control the flow of information in normal and cancer cells: DNA, RNA, and Proteins. The first portion of the volume describes the use of microarrays and next generation sequencing for genome-wide analysis of genetic and epigenetic variation. The next portion provides an overview of these technologies in the study of gene expression at the RNA level. The final section details the use of mass spectrometry and tissue microarrays for high-throughput and parallel analysis of proteins. As these technologies are deployed in cancer research, and the analytical approaches for interpretation of the resulting data mature, we will greatly increase our fundamental understanding of carcinogenesis and be able to translate this understanding for development of biomarkers and therapeutic strategies in the dawning era of individualized medicine.

Contents

Contributors

Mohamad Abbani
Cedars-Sinai Medical Center,
Los Angeles, CA, USA;
Department of Chemistry and Biochemistry,
University of California at Los Angeles,
Los Angeles, CA, USA
abbanim@cshs.org

Esteban Ballestar
Cancer Epigenetics Group,
Spanish National Cancer Research Centre (CNIO),
Melchor Fernández Almagro 3, 28029, Madrid, Spain
eballestar@cnio.es

Yidong Chen
National Cancer Institute,
Bethesda, MD, USA
yidong@mail.nih.gov

Toby Cornish
Department of Pathology,
Johns Hopkins University School of Medicine,
Baltimore, MD, USA
tcornis2@jhmi.edu

Angelo De Marzo
Department of Pathology,
Johns Hopkins University School of Medicine,
Baltimore, MD, USA
ademarz@jhmi.edu

Manel Esteller
Cancer Epigenetics Group,
Spanish National Cancer Research Centre (CNIO),
Melchor Fernandez Almagro 328029, Madrid, Spain
mesteller@cnio.es

Leroy Hood
Institute for Systems Biology,
Seattle, WA, USA
lhood@systemsbiology.org

William B. Isaacs
Sidney Kimmel Comprehensive Cancer Center,
Brady Urological Institute,
Johns Hopkins University School of Medicine,
Baltimore, MD, USA
wisaacs@jhmi.edu

Biaoyang Lin
Department of Urology,
University of Washington, Seattle, WA, USA;
Zhejiang-California International Nanosystems Institute,
Hangzhou, China
bylin@u.washington.edu

Jun Luo
Department of Urology,
Johns Hopkins University School of Medicine,
Baltimore, MD, USA
jluo1@jhmi.edu

Parag Mallick
Center for Applied Molecular Medicine,
University of Southern California,
Los Angeles, CA, USA;
Department of Chemistry and Biochemistry,
University of California at Los Angeles,
Los Angeles, CA, USA
parag@ucla.edu

William G. Nelson
Sidney Kimmel Comprehensive Cancer Center,
Johns Hopkins University School of Medicine,
Baltimore, MD, USA
bnelson@jhmi.edu

Maryann Vogelsang
Cedars-Sinai Medical Center,
Los Angeles, CA, USA;
Department of Chemistry and Biochemistry,
University of California at Los Angeles,
Los Angeles, CA, USA
maryann.vogelsang@cshs.org

Jeremy Wechsler
University of Washington,
Seattle, WA, USA

Sarah J. Wheelan
Sidney Kimmel Comprehensive Cancer Center,
Johns Hopkins University School of Medicine,
Baltimore, MD, USA
swheelan@jhmi.edu

Srinivasan Yegnasubramanian
Sidney Kimmel Comprehensive Cancer Center,
Johns Hopkins University School of Medicine,
Baltimore, MD, USA
syegnasu@jhmi.edu

Chapter 1
Genome-Scale Analysis of Data from High-Throughput Technologies

Sarah J. Wheelan

Abstract Few technical advances have excited such a broad spectrum of basic and clinical scientists as high-throughput technologies (microarrays and sequencing). Having learned in training that somewhere in the genome lies the key to just about any phenotype, scientists are fast joining the movement to decrease cost and improve access to these technologies. Generating enormous amounts of high-dimensional data brings certain challenges, and many researchers are turning even further from their training to collaborate with computer scientists and biostatisticians, who are equally excited to analyze these promising datasets. As new and truly interdisciplinary teams are created, we are seeing major advances; the current environment is exciting for all involved. Technology has brought entire scientific fields to the brink of discovery before, and will again, and thus the overall enthusiasm must be tempered by the fact that new technology brings new problems and new artifacts that we have not seen before. We can circumvent some of these by paying careful attention to experimental design, staying mindful of the complexities of the underlying biology, and by soliciting assistance from analysts versed in high-dimensional data.

1.1 The Genomic Scale

The landscape of the human genome, with a haploid size of about 3 gigabases, becomes more complex as our understanding of it grows, and the genomics community is likely still unaware of its greatest treasures and surprises. A cell's DNA sequence and epigenome direct its growth, differentiation, and gene expression.

S.J. Wheelan (✉)
Sidney Kimmel Comprehensive Cancer Center, Johns Hopkins University School of Medicine, Baltimore, MD, USA
e-mail: swheelan@jhmi.edu

S. Yegnasubramanian and W.B. Isaacs (eds.), *Modern Molecular Biology*, Applied Bioinformatics and Biostatistics in Cancer Research,
DOI 10.1007/978-0-387-69745-1_1,

If we better understand how the sequence and associated modifications work, it holds incredible promise for oncology, medicine, genetics, and all of biology. Individualized medicine is an early venture that aims to provide therapies tailor-made for a patient's particular genome.

Cancer cells have multiple and multifaceted changes in their genomes, and these alterations and their ensuing phenotypic effects involve nearly every known genetic process, including tissue growth and differentiation, small and large DNA mutations, methylation effects, RNA-mediated effects, and gene regulation. For this reason, a genomic approach to cancer seems appropriate, though challenging. Fortunately, newer molecular biology tools promise to yield a very broad understanding of any specific genome.

Molecular biology is in transition. After decades of perfecting small-scale, single- or several-molecule experiments whose results are generally qualitative and interpretable by eye, scientists are confronted with a sudden shift in the culture of the field as new technology produces, in a matter of days, quantities of data that earlier generations of researchers would not have seen in an entire career (Shendure and Ji 2008). Simply storing and transferring these data is an unsolved problem (current methods include mailing large hard drives and carrying stored data between buildings in carts) (Marshall 2008), and analytical methods are in early stages. Even when data analysis is complete, inferring biological meaning from extremely large datasets is difficult and haphazard, given current databases and algorithms.

For example, a transcription factor that is thought to positively regulate a pathway, given prior experimental data, may, when analyzed in the context of the entire genome, have much more complicated effects, some of which may even cancel out the effects of the pathway that the factor was thought to promote. A biologist's intuition is challenged by these data, and results obtained from using methods designed for smaller-scale analyses may be incomplete or misleading.

An even more fundamental difference is that experiments are becoming data-driven, rather than hypothesis-driven, meaning that a result was not generated within the framework of a specific question, but instead as part of a broad-based scan using a given technology. This can be a problem for statistical analysis, as a more classical hypothesis-driven experiment asks a single question and the results can be interpreted as supportive, contradictory, or statistically inconclusive; while a data-driven experiment simply produces lots of data, which must be sifted through in a search for interesting trends and biases. Distinguishing signal from noise on this scale is difficult without a solid understanding of the biological and technical issues at hand, and scientists can happily cherry-pick a promising result without understanding its overall relevance.

Genome-scale experiments must be conceived, performed, and analyzed with these complexities and uncertainties in mind, and in close collaboration with analytical specialists. With careful choices of experimental and analytical techniques, an investigator can make significant progress and make good use of expensive high-throughput techniques.

Many static and dynamic features of a genome can already be examined through high-throughput technology, and new techniques promise even more power and flexibility. Features that can be queried include those defined by sequence characteristics (exons, introns, pseudogenes, binding sites, repetitive sequence, transposons, retrotransposons, endogenous retroviruses, centromeres, telomeres), those less well-defined by sequence alone (miRNA, promoters, CpG islands, noncoding RNAs), and those for which the sequence composition, if any, is unknown (nucleosome-associated sites, matrix attachment regions). In addition, epigenetic phenomena such as methylation critically affect a cell's function, and recent advances allow better definition of this class of features. This is a subset of what is understood, and future major discoveries will bring to light genomic features that we cannot imagine yet.

As genome sequencing data continue to accumulate, scientists can turn to evolutionary methods and phylogenetic analysis to search for sequences that are functionally important in cancer, and that may not be among the elements that are easily identified by sequence alone. Having only a single representative genome for most of the very few fully sequenced organisms limits the power of evolutionary techniques; however, significantly conserved non-genic sequences found so far have had regulatory roles that would not have been uncovered otherwise (Dermitzakis et al. 2003; El-Mogharbel et al. 2007).

Within the human population, the full variation among genomes is not well characterized (Weiss 1998). The number of single bases that are different from one person to another is likely to be rather small, though that calculation was made from a limited number of genomes, but larger-scale changes like copy number variation, which are much more difficult to detect, seem to be very common. Any given change may have effects ranging from devastating to undetectable, and the phenotypic effects of a mutation often cannot be predicted from examining the genome sequence; for example, cystic fibrosis and achondroplasia are serious diseases resulting each from a single nucleotide change, whereas most identified SNPs are apparently silent. Only when large, population-wide projects (for example, the 1,000 genomes project) are completed will we get a better picture of the genetic variation in our species, and its significance (Kaiser 2008).

Complicating this, not all somatic cells in a single person possess exactly the same genome. Aside from developmentally timed rearrangements, such as recombination in cells in the immune system, cells gain mutations over time because of environmental damage and mistakes made in DNA replication. It is possible and even likely that some mutations that occur during somatic cell growth and division are not detrimental and are retained. Epigenetic variations also distinguish cell types and are the subject of intense study. These changes do not affect the primary DNA sequence but profoundly affect gene expression and regulation, and are one source of the unexpected phenotypic variation often seen even in nondiseased whole-tissue samples.

To make things more difficult, looking for tiny variations in populations is problematic with current error-prone and generally coarse technology. Microarrays query only a subset of the genome and will therefore miss any variation falling in sequences between probes, and microarray data only rarely allow a researcher to distinguish between very slight sequence variants, as typical probes are not very sensitive to single base differences. Sequencing does not generally cover the genome evenly, and the technique's error rate complicates definitive identification of rare polymorphisms.

As a tumor is a complex structure that may be more of a population than a tissue, looking at cancers using genome-wide techniques is promising but difficult. Tumor cells are likely to possess slightly different genomes and the variations have unknown effects, and a scientist must obtain significant amounts of data in order to have confidence that the observed variations are real and not experimental errors.

1.2 Different Experimental Designs Focus on Different Biological Processes

Genomic experiments can either query the genome in an unbiased way, without respect to known biological phenomena (tiling array, or whole genome sequencing), or can focus on a selected subset of the genome, such as expressed sequences, sequences bound to a known factor, or sequences enriched in some other biologically meaningful way. While examining the entire genome clearly seems to be the superior approach, it is expensive and the data can be difficult to interpret. If the hypothesis allows, a targeted approach can be more economical and the data are easier to place into a statistical framework.

Resequencing an entire genome may be necessary for diseases such as cancer, in which the affected cells have likely undergone multiple and major genetic events, and sequencing only parts of their genomes may give a very incomplete picture. When a biological process is known to affect regions of the genome and these regions can be isolated by chromatin immunoprecipitation, PCR, sequence capture, or any other molecular biology method, sequencing these regions only is the best way to gain information about what sequences and molecules are involved in the process of interest.

One obvious subset of the genome is the transcribed sequences. Using various methods to capture mRNA, the expression patterns of different cells and tissues in different stages of development or disease can be compared (Gnirke et al. 2009). While the statistical analysis of these data can be complex, recent experiments have identified interesting variations in expression, including an unexpectedly wide range of tissue-specific expression of alternative spliceoforms, and have revealed transcription from a much broader set of sequences than previously estimated (Pan et al. 2008).

A caveat when dealing with mRNA measurements is that there is a surprising variety of splicing products for nearly every gene. If measurements are made in only a few areas of the coding sequence (for example, from a microarray with a few probes per gene), the expression pattern of the gene may be completely mischaracterized (for example, if the probes query exons that only participate in a rarer spliceoform). Recent methods have focused on splice sites in an attempt to remedy this problem, but these methods (for example, exon arrays) still cannot measure completely unanticipated products, as when cryptic exons are included or cryptic splice sites are used. Short read sequencing is a difficult technique to apply to this problem, since it produces so little sequence on either side of the splice junction that mapping the two exons is difficult and error-prone. Longer reads may help with these cases so that the splice junctions can be properly mapped (Trapnell et al. 2009).

Other genomic subsets include transposons, which can be amplified by various forms of targeted PCR, as well as a number of sequence- and structure-defined regions such as methylated blocks, transcription factor-bound sequences, and sequences associated with chromatin structures or the nuclear matrix. Protocols involving antibody recognition are commonly used to target and amplify these DNA regions, and statistical techniques have been developed to analyze these sequences when bound to an array (ChIP-chip) or sequenced directly (ChIP-seq). These approaches can reveal large-scale regulatory and mutational changes in a cancer genome that may not be obvious if the entire genome is sequenced (Wheelan et al. 2006).

1.3 Analytic Approaches Fall into Three Categories

Once sequencing or microarray data are obtained, analysis can proceed along three paths.

Mutations and regulatory changes may fall in regions that have been studied already in wet lab settings or are otherwise well characterized. Several techniques, such as gene set enrichment analysis, genome wide association, and network and pathway methods, already exist for analysis of these data.

Often, the changes that are identified in a cancer cell that do not exist in its matched samples do not fall within a genomic region that is thoroughly understood. In these cases, evolution-based techniques, such as cross-species alignment to organisms that have been better studied, can take advantage of results of experiments done in other species.

Finally, if there is no direct functional information or a strong cross-species correlation, the mutations can be analyzed at a sequence level. This involves looking for motifs, binding sites, changes in complexity or composition, and gene and spliceoform prediction. This step should be performed regardless of what is learned at other levels of analysis, as long-range DNA and protein interactions that act on the sequence level may otherwise be hard to detect.

1.4 Databases and Sequence Repositories Play a Key Role in Modern Genomics Research

Analyzing microarray and sequencing data requires comparing the data to databases of annotated sequences, pathways, and functions. The statistical questions posed by these analyses are formidable and not completely solved; a major problem is that no database is truly independent from any others, so that the database queries are not always statistically independent.

Current molecular databases are numerous and include those which catalog nucleic acid and protein sequences, molecules with functional annotations, assembled functional pathways, experimental results such as microarray output, and genes linked to human diseases.

GenBank, the sequence repository now maintained at the National Institutes of Health, was started in 1982. Originally holding just nucleic acid sequences, it was distributed in paper form, then on diskettes and CD-ROM (articles included a phone number to call to obtain the CD), and online. After a massive and ever-accelerating increase in sequence input, GenBank has far surpassed the hundred billion base mark, has spawned several sister divisions, and now contains more information than can ever be analyzed by currently existing tools (Strasser 2008; Benson et al. 2009).

Now that disk space is relatively cheap and web interfaces are easy to make, databases have proliferated to where any field in medicine or genetics, no matter how specialized, sports at least one or two custom databases. Unfortunately, data are circulated among databases so freely that different databases rarely contain completely different data, and errors tend to propagate, even among manually curated databases. Databases are so interdependent that using more than one in an analysis may introduce unpredictable biases, as the supposedly independent confirmation from the second database may suffer from the same errors as the first result (Jones et al. 2007).

A good example of database problems comes from a 2004 paper that traced systematic and widespread gene name errors to problems in data formatting by researchers (Zeeberg et al. 2004). Unfortunately, due to the way the errors occurred, fixing them is impossible.

1.5 Sequencing

DNA sequencing has evolved from a painstaking manual lab method that produced a few hundred bases, did not require much downstream analysis, and answered a single question, to a painstaking semi-automated contract lab or core method that yields millions to billions of bases, requires extensive analysis, and usually raises many questions.

Enumerating and discussing the various and interesting methods and chemistries of modern sequencing methods is not useful at the moment, as the landscape is changing rapidly. Current high-throughput next generation sequencing platforms include those commercialized by Illumina (http://www.illumina.com), Life Technologies/Applied Biosystems (http://www.appliedbiosystems.com), Roche (http://www.454.com), Helicos (http://www.helicosbio.com), and Complete Genomics (http://www.completegenomics.com). While each of these platforms feature different sequencing chemistries, imaging formats, library preparation strategies, etc., the each generate millions of sequencing reads, spanning in size from ten to a few hundred base pairs depending on the platform, in each run. Despite this diversity, the basic experimental design and types of output are useful to think about.

Two broad classes of sequencing experiments are whole genome and targeted. The first generates sequences that somewhat evenly cover the genome, and targeted sequencing generates data that are clumped to some degree or another. Clearly, these are simplifications, as a whole genome experiment does not produce perfectly even coverage, and a targeted experiment often produces sequence reads speckling the intervals between the intended targets, which appear as clumps of reads.

Whole genome sequencing simply attempts to sequence every base in the input genome at enough depth that polymorphisms can be separated from sequencing error.

Targeted sequencing methods include ChIP-based approaches, capture techniques, and many more. Any molecular biology method that produces a set of DNA or RNA sequences thought to share some biological characteristic can be used to create samples for targeted sequencing. Targeted sequencing approaches are especially useful because they pare down the range of genomic loci expected in the output, so the alignment phase can often be simplified.

1.6 Alignment, Mapping, and Assembly

Microarray experiments involve binding sample sequences to probes with known sequence and mapping information. The amount of binding to each probe must be determined, but the probe sequences already have known genomic positions. When high-throughput sequencing is the experimental method used, the data must be aligned and mapped before downstream analysis can begin.

A text file of 100 million short sequence reads is not helpful without annotation. For resequencing applications, the reads can be aligned to the known sequence and mapped relative to known sequence features, and their polymorphisms can be catalogued against the reference sequence.

Alignment of millions of short sequences to a gigabase-scale genome is technically challenging. A tool like BLAST, which is very flexible and can adapt to many situations, is far too slow to be useful in this procedure, although it does allow for mismatches, a feature that causes problems in many other programs. There are dozens of alignment programs created expressly to align short read sequences to a reference genome. Each has different options and runs at different speeds. The fastest can align millions of reads in seconds, though when more advanced options are used (such as allowing mismatches), most programs get slower. Most of the programs do take advantage of the quality scores that the instrument creates for each read, allowing mismatches preferably at lower quality positions (Trapnell and Salzberg 2009).

When a new genome is sequenced, the challenges are much different. Depending on the availability of close evolutionary neighbors, known reference sequences may still be quite helpful in assembling the unknown genome. A technique called gene-boosted assembly was recently reported (Salzberg et al. 2008) and seems promising for the assembly of new or highly mutable genomes. This technique takes advantage of conserved gene order to help piece together reads that span recognizable genes and unconserved flanking regions. Other techniques, many inspired by recent interest in metagenomics, are promising as well (Pop 2009).

Cancer genome analysis may be the most challenging application of high-throughput sequencing, as it combines metagenomics with a high mutation rate. Investigating tumor specimens or cancer cell lines with short read sequencing falls somewhere between resequencing and de novo sequencing in terms of technique. While the starting genome is already known, cancer cells often harbor severe and dramatic amplifications, deletions, point mutations, and gross changes such as translocations. Translocations may be especially difficult to detect, as a new junction is created that is not present in the reference genome. In this case, successful alignment requires a translocation junction to be flanked, in a single sequence read, by sufficiently long pieces of the chromosomes from either side of the junction so that a believable alignment can be generated from each piece; otherwise, a fragment that maps partially to two chromosomes will be discarded.

1.7 Microarrays

Sequencing is one of the newer high-throughput techniques, but many others exist and can be just as efficient, often at a lower cost, depending on the problem under study. Microarrays are glass slides or other solid substrates to which single-stranded DNA molecules have been affixed, in a known configuration (Wheelan et al. 2008).

DNA (or RNA) is isolated from a sample and hybridized to the array after labeling with a fluorescent dye or other easily detectable (and quantifiable) marker. Theoretically, the intensity of the signal at each spot on the array is proportional to the amount of its complementary DNA sequence in the original sample.

Microarrays are popular, so that even the newest and highest-capacity varieties can be ordered online with custom sequences, for a cost that is reasonable for most experimental labs. Microarrays can have millions of features (spots of DNA, or probes, with different sequences) and some slides are configured with several smaller, identical arrays that can be hybridized to different samples simultaneously, which reduces signal fluctuations caused by technical variations.

Several factors confound interpretation of microarray data. First is the probe effect: some probes bind their complements better than others, and if all DNA in the sample is at equal concentration, those probes will have a more intense signal. Another difficulty in analyzing microarray data is that some areas of the slide may be more or less intense than others, because of the way the array was handled, or sometimes because of the way the array was printed, even though the probes in the different areas may be complementary to target DNA that is at similar concentrations in the sample. There are robust and straightforward statistical techniques for handling these variations, but the analysis must be done carefully to avoid introducing even more false signals. Additionally, the subtle differences in the way that different technicians handle arrays are also important, as is the calibration of the scanner. In fact, the data vary in predictable ways from technician to technician and day to day (batch effects) such that a statistician can often detect which arrays were run by which technician, and in which order they were processed. A similar phenomenon exists for sequencing data, but has not been investigated as thoroughly.

As with sequencing, microarrays can be used in either targeted or unbiased experiments. A tiling array is a relatively unbiased design for a microarray in which the probe sequences are chosen without regard to genomic features and are spaced as evenly as possible across the genome. Targeted arrays contain probes from specific functional or structural subsets of the genome, and are intended to answer narrower questions than tiling arrays.

Tiling arrays, by design, are able to uncover biological phenomena that targeted arrays cannot, though targeted arrays can be simpler to analyze. For example, looking at transcription patterns of known genes is very different than looking at transcription of random sequences throughout the genome; the latter types of experiments, done previously using tiling arrays, suggested that a much larger fraction of the genome is transcribed than was previously suspected (Kapranov et al. 2002, 2007; Cheng et al. 2005). The significance of this finding is unknown and difficult to test, which is one downside of looking at the behavior of unfamiliar sequences.

1.8 Experimental Design Considerations

An experiment that uses high-throughput genomics methods must be conceived and designed very differently from a more traditional biological or clinical assay, not only because it is expensive but because its output is easily misunderstood.

Most importantly, if there will be a statistician, biostatistician, or other type of analyst involved in the experiment, the individual should be identified early and should help create the experimental design. It is too easy to overlook this simple consultation and end up with results that can never be significant due to flaws in the approach. Moreover, the analyst who has experience with genome-wide approaches will have many suggestions for different techniques and methods that another scientist might not have considered.

Follow-up studies can use better understood and cheaper methods (PCR and capillary sequencing, for example). An investigator must define the goals and expectations of each stage in the experiment to take advantage of the huge variety of techniques now available. Because it is unlikely that a scientist will have the resources to independently confirm every promising data point, it is necessary to prioritize, which should often be done more skillfully than taking the top ten from a long list of statistically significant scores.

Finally, there is the question of what to do with all these data. The data are valuable, as the experiments were costly and time-consuming, but the data themselves occupy a lot of disk space and may or may not ever be revisited. Databases are overflowing (the size of GenBank continues to skyrocket) and data repositories such as GEO are generally used only when results are to be published, so they represent a skewed version of a lab's output. Simply transferring large datasets can take hours or more, even with high-capacity connections. These issues are becoming more critical and may soon be the limiting factor in scientific discovery.

1.9 Conclusions

Genome-scale experiments hold great promise for cancer researchers. With appropriate planning and choice of technology, scientists can design experiments that can yield usable and significant data. These approaches should be considered hypothesis generators at the moment, as biological and clinical knowledge is not sophisticated enough and analytical techniques not advanced enough for a large-scale dataset to be its own result rather than a source of information for further studies.

References

Benson DA, Karsch-Mizrachi I, Lipman DJ, Ostell J, Sayers EW (2009) GenBank. Nucleic Acids Res 37:D26–D31

Cheng J, Kapranov P, Drenkow J, Dike S, Brubaker S, Patel S, Long J, Stern D, Tammana H, Helt G, Sementchenko V, Piccolboni A, Bekiranov S, Bailey DK, Ganesh M, Ghosh S, Bell I, Gerhard DS, Gingeras TR (2005) Transcriptional maps of 10 human chromosomes at 5-nucleotide resolution. Science 308:1149–1154

Dermitzakis ET, Reymond A, Scamuffa N, Ucla C, Kirkness E, Rossier C, Antonarakis SE (2003) Evolutionary discrimination of mammalian conserved non-genic sequences (CNGs). Science 302:1033–1035

El-Mogharbel N, Wakefield M, Deakin JE, Tsend-Ayush E, Grutzner F, Alsop A, Ezaz T, Marshall Graves JA (2007) DMRT gene cluster analysis in the platypus: new insights into genomic organization and regulatory regions. Genomics 89:10–21

Gnirke A, Melnikov A, Maguire J, Rogov P, LeProust EM, Brockman W, Fennell T, Giannoukos G, Fisher S, Russ C, Gabriel S, Jaffe DB, Lander ES, Nusbaum C (2009) Solution hybrid selection with ultra-long oligonucleotides for massively parallel targeted sequencing. Nat Biotechnol 27:182–189

Jones CE, Brown AL, Baumann U (2007) Estimating the annotation error rate of curated GO database sequence annotations. BMC Bioinformatics 8:170

Kaiser J (2008) DNA sequencing. A plan to capture human diversity in 1000 genomes. Science 319:395

Kapranov P, Cawley SE, Drenkow J, Bekiranov S, Strausberg RL, Fodor SP, Gingeras TR (2002) Large-scale transcriptional activity in chromosomes 21 and 22. Science 296:916–919

Kapranov P, Cheng J, Dike S, Nix DA, Duttagupta R, Willingham AT, Stadler PF, Hertel J, Hackermuller J, Hofacker IL, Bell I, Cheung E, Drenkow J, Dumais E, Patel S, Helt G, Ganesh M, Ghosh S, Piccolboni A, Sementchenko V, Tammana H, Gingeras TR (2007) RNA maps reveal new RNA classes and a possible function for pervasive transcription. Science 316:1484–1488

Marshall A (2008) Prepare for the deluge. Nat Biotechnol 26:1099

Pan Q, Shai O, Lee LJ, Frey BJ, Blencowe BJ (2008) Deep surveying of alternative splicing complexity in the human transcriptome by high-throughput sequencing. Nat Genet 40:1413–1415

Pop M (2009) Genome assembly reborn: recent computational challenges. Brief Bioinform 10:354–366

Salzberg SL, Sommer DD, Puiu D, Lee VT (2008) Gene-boosted assembly of a novel bacterial genome from very short reads. PLoS Comput Biol 4:e1000186

Shendure J, Ji H (2008) Next-generation DNA sequencing. Nat Biotechnol 26:1135–1145

Strasser BJ (2008) Genetics. GenBank – Natural history in the 21st Century? Science 322:537–538

Trapnell C, Salzberg SL (2009) How to map billions of short reads onto genomes. Nat Biotechnol 27:455–457

Trapnell C, Pachter L, Salzberg SL (2009) TopHat: discovering splice junctions with RNA-Seq. Bioinformatics 25:1105–1111

Weiss KM (1998) In search of human variation. Genome Res 8:691–697

Wheelan SJ, Scheifele LZ, Martinez-Murillo F, Irizarry RA, Boeke JD (2006) Transposon insertion site profiling chip (TIP-chip). Proc Natl Acad Sci U S A 103:17632–17637

Wheelan SJ, Martinez Murillo F, Boeke JD (2008) The incredible shrinking world of DNA microarrays. Mol Biosyst 4:726–732

Zeeberg BR, Riss J, Kane DW, Bussey KJ, Uchio E, Linehan WM, Barrett JC, Weinstein JN (2004) Mistaken identifiers: gene name errors can be introduced inadvertently when using Excel in bioinformatics. BMC Bioinformatics 5:80

Chapter 2
Analysis of Inherited and Acquired Genetic Variation

Srinivasan Yegnasubramanian and William B. Isaacs

Abstract Cancer is a genetic disease. While this statement is accurate, its simplicity betrays the underlying complexity of the genetic alterations. The first layer of complexity comes from the contribution of inherited vs. acquired genetic variation in cancer initiation and disease progression. Several other layers of complexity come from the myriad types of genetic variation, such as point mutations, amplifications, deletions, translocations, inversions, etc., that have been directly and causally linked with human malignancies. Additional layers of complexity arise from the interactions of genetic alterations with environmental exposures to drive further genetic as well as epigenetic alterations. In this chapter, we will introduce the modern tools available to decipher the complex genetic variation contributing to the initiation and progression of cancer. In Chaps. 3 and 4, we will introduce the modern tools available for understanding epigenetic alterations, such as heritable patterns in DNA/chromatin and protein interactions and DNA methylation.

2.1 Introduction

It is clear that cancer is a complex disease that ultimately arises from molecular genetic alterations (Hanahan and Weinberg 2000). A number of these alterations may be inherited and lead to increased susceptibility (Houlston and Peto 1996; Lichtenstein et al. 2000), and many other alterations are acquired through life, perhaps due to behaviors, exposures and other environmental factors. The inherited genetic traits may also influence the susceptibility to acquire various cancer associated

S. Yegnasubramanian (✉)
Sidney Kimmel Comprehensive Cancer Center, Johns Hopkins University, School of Medicine, Baltimore, MD, USA
e-mail: syegnasu@jhmi.edu

S. Yegnasubramanian and W.B. Isaacs (eds.), *Modern Molecular Biology*, Applied Bioinformatics and Biostatistics in Cancer Research,
DOI 10.1007/978-0-387-69745-1_2,

genetic alterations through life. In this chapter, we will first introduce various concepts in cancer genetics, and then discuss the modern high-dimensional approaches used to understand the genetic alterations that drive cancer initiation and progression.

2.1.1 Oncogenes and Tumor Suppressor Genes

In the classical view, normal human somatic cells contain mechanisms that, in certain contexts, promote proliferation, for example in response to tissue injury or during development and growth, and also mechanisms that keep this proliferation in check, for example genes that promote senescence or apoptosis in response to cellular or genomic injuries that cannot be completely repaired (Hanahan and Weinberg 2000). In a simplistic sense, cancers are typically thought to arise by constitutive activation of genes, called oncogenes, that promote growth and proliferation, usually accompanied by inactivation of genes, called tumor suppressor genes, that normally regulate and limit this proliferation. In many situations, it is known that activation of an oncogene without loss of relevant tumor suppressor genes can lead to activation of cellular mechanisms that promote senescence or apoptosis, resulting in suppression of tumor formation (Braig and Schmitt 2006). After examining the rate of development of bilateral and unilateral retinoblastoma in familial and sporadic cases, Al Knudson inferred that tumor formation required the acquisition of two distinct hits (Knudson 1971). In familial cases, the first of these hits was already inherited and the second was acquired during life. In sporadic cases, both hits had to be acquired during life. These results, taken across a cohort of individuals with familial and sporadic retinoblastoma, combined with estimates on mutational frequencies could be used to predict a model in which two mutations or hits in an individual cell were needed to give rise to a clonal tumor outgrowth, later referred to as the two-hit hypothesis. Indeed, later studies identified the causal mutations in the retinoblastoma gene (RB), and confirmed that both alleles were inactivated by genetic processes such as mutation or copy number loss or even by epigenetic processes such as promoter DNA hypermethylation (see Chap. 4 for discussion of DNA methylation) in retinoblastoma tumors, both in familial and sporadic cases (Jones and Laird 1999; Knudson 2001). This inference became known as the Knudson two-hit hypothesis for inactivation of tumor suppressor genes, and has remained one of the central tenets of cancer biology. Much of cancer molecular genetics has involved identification of the causal mutations that activate a host of oncogenes and inactivate tumor suppressor genes.

2.1.2 Types of Genetic Variation and Alteration in Human Cancer

There are many types of mutations that can cause inherited or acquired susceptibility to carcinogenesis and disease progression. Point mutations involve the alteration of a single nucleotide by substitution with another nucleotide, deletion, or insertion. When these occur in non-protein coding regions of the genome, such changes have unclear ramifications. However, in protein coding sequences, these changes can be silent, i.e. result in a change that does not alter amino acid sequence in the protein (due to the redundancy in the genetic code of triplet codons), or can be non-synonymous. Among the non-synonymous mutations, both missense mutations, resulting in a change of one amino acid to another, or nonsense mutations, resulting in a premature stop codon and truncation, are possible. Insertions and deletions can lead to frame-shift mutations completely altering the downstream amino acid sequence or resulting in protein truncation via generation of premature stop codons. Point mutations can often be caused by polymerase errors or by genotoxins that adduct to DNA bases and alter fidelity of replication of those bases in characteristic ways. Specific carcinogens have been observed to cause a specific spectrum of mutations. Mutations that result in the change of a purine base to purine (A to G or vice versa) or a pyrimidine to pyrimidine (C to T or vice versa) are called transition mutations. The transition of C to T can occur, for instance, by the spontaneous deamination of methylcytosine to thymine. Mutations that result in the change of a purine to pyrimidine or vice versa are referred to as transversion mutations and are less common than transition mutations. Larger scale mutations include amplifications, resulting in multiple copies of specific genomic segments, deletions, resulting in loss of specific genomic segments, chromosomal translocation, leading to rearrangement and juxtaposition of genomic segments from nonhomologous chromosomes, and inversions, leading to a reversal of the orientation of a chromosomal segment. Translocations and inversions can occur in a balanced fashion, preserving copy number, or unbalanced fashion, leading to loss or gain of regions near the affected region. Mutations can be further classified according to functional parameters (loss- or gain-of-function, dominant negative, lethal, etc.) depending on the effect of the mutation on a specific gene or cell. Each of these mutations can be interrogated by specialized approaches, as will be described below, depending on the types of libraries prepared prior to analysis.

2.1.3 Familial Cancer Syndromes and Link to Sporadic Cancers Affecting the Same Organ Sites

Just as in the case of familial retinoblastoma, mutations found to be causal in many familial cancer syndromes are also found in sporadic cases of cancers arising in the same organ system (Kinzler and Vogelstein 1996). For example, in the familial colon cancer syndromes adenomatous polyposis coli (APC) and hereditary nonpolyposis colorectal cancer (HNPCC), causal mutations in one copy of the APC gene and in mismatch repair genes respectively have been found in the germline of affected families. The second copy then becomes mutated in individual cells that then produce clonal outgrowths to form polyps and/or cancers. Interestingly, a large fraction of individuals developing sporadic colon cancer show biallelic loss of APC and/or mismatch repair genes. Thus, the study of hereditary cancers by using classical genetic approaches to narrow down regions associated with disease in affected individuals followed by candidate gene approaches is a classical means for identification of causal gene mutations that can lead even to sporadic disease.

2.1.4 *Inherited Susceptibility to Common Sporadic Cancers and Role of Environmental/Lifestyle Factors in Modifying Risks*

The age-old questions of nature vs. nurture and inherited vs. acquired traits are highly relevant to cancer research. Most cases of cancer do not show clear cut patterns of Mendelian inheritance. Nonetheless, many common sporadic cancers appear to have family history as a risk factor in development of disease (Houlston and Peto 1996). However, since individuals from families may have common environmental exposures in addition to common inherited traits, it is difficult to determine what role inherited genetic susceptibility alone or environmental/lifestyle factors alone play in carcinogenesis. To clarify such confounding issues, epidemiologists and cancer researchers have used twin studies to compare the concordance of development of cancer between twins in monozygotic vs. dizygotic pairs (Lichtenstein et al. 2000). Since monozygotic twins would have identical genetic make-up, they should have an equal genetic contribution to development of disease. Additionally, the fact that both monozygotic and dizygotic twins would be expected to share environmental exposures between twins to a similar degree helps to control for the effect of environmental exposures. As an example, using such twin study designs, it has been estimated that 42% of the risk of acquiring prostate cancer is genetically determined (Lichtenstein et al. 2000), and this rate is among the highest estimates for all cancers. This suggests that environmental/lifestyle factors make up the remaining risk factors. Some hints regarding these environmental/lifestyle contributions have been obtained from epidemiological studies examining the rates of developing cancer in different geographic regions. Again using prostate cancer as an example, epidemiological studies have shown a much stronger prostate cancer incidence, prevalence, and mortality among western, industrialized nations compared to eastern nations (Hsing et al. 2000). However, since some of these differences may be due to ethnic and genetic differences between men in these different geographic regions, it was difficult to assess the contribution of environmental/lifestyle factors alone. To tease this apart, studies examining the rates of prostate cancer incidence, prevalence and mortality in immigrants from eastern nations that recently emigrated vs. those that had settled in western industrialized nations for much longer periods, likely adopting more western cultural/dietary/lifestyle behaviors than their recently emigrated counterparts, showed a higher risk for prostate cancer development and mortality, approaching the rates of those seen in men native to western industrialized nations (Haenszel and Kurihara 1968; Shimizu et al. 1991; Whittemore et al. 1995). These studies provide more evidence that factors in the environment and lifestyle can contribute to the acquisition of genetic alterations during life that contribute to cancer risk. Taken together, the body of evidence suggests that it is of prime importance to understand the inherited genetic determinants of cancer susceptibility as well as to elucidate the acquired genetic alterations that lead to transformation of individual cells in the formation and progression of cancer.

2.2 Use of Microarrays for Genome-Wide Analysis of Genetic Variation/Mutation

While the first microarrays were used for analysis of gene expression, DNA microarrays allowing analysis of alterations to genomic sequence have emerged as powerful tools for the study of cancer-related genomic alterations. Prior to the advent of microarrays, much of the early work examining genomic alterations in cancer cells came from karyotypic analysis. In these analyses, cells obtained from a cancer or tissue specimen would be isolated in primary culture, and metaphase spreads would be obtained via treatment of these cells by a solution of colchicine to inhibit the spindles and arrest mitosis. These metaphase spreads would then be stained (e.g. with Giemsa), and using the size and pattern of banding (Drets and Shaw 1971), individual chromosomes, and rearrangements or alterations in those chromosomes (Rowley 1973) could be identified for a handful of metaphase spreads. Spectral karyotyping (SKY) is a variation of this technique that uses labeling with probes for each chromosome labeled with different proportions of fluorescent dyes combined with spectral imaging to assign different colors to each chromosome instead of staining with Giemsa (Schrock et al. 1996). Karyotyping analysis is usually limited to analysis of cells that could be cultured and processed to obtain metaphase spreads. The advent of fluorescence in situ hybridization (FISH) (Bauman et al. 1980) and chromosome painting (Cremer et al. 1988; Lichter et al. 1988; Pinkel et al. 1988) allowed more sophisticated microscopic examination of genetic alterations in interphase nuclei. Such analyses are relatively labor intensive, can take significant amount of time, and have very low resolution compared to other modern analytical approaches.

2.2.1 *Comparative Genomic Hybridization*

The first approaches for comparative genomic hybridization (CGH) (Kallioniemi et al. 1992; du Manoir et al. 1993) represented a significant extension of karyotype analysis. The goals of these experiments were to compare the relative chromosomal copy number of a sample, such as tumor genomic DNA, against a reference, such as a matched normal genomic DNA sample. Normal metaphase spreads are made as the substrate to which hybridization is carried out. Then, genomic DNA from the tumor sample is labeled with one fluorescent dye, and genomic DNA from the normal sample is labeled with a different fluorescent dye. These two labeled DNAs are then mixed in equal proportions, along with unlabeled cot-1 DNA, which is made up of highly repetitive elements from the human genome (allowing suppression of hybridization of repetitive DNA), and hybridized to the metaphase spread. The relative fluorescence intensity of one color to the other along the length of the chromosome would provide the relative copy number of the sample against the reference. This method was very useful for determination of large amplifications and deletions affecting at least one to three MBps. Major advances to this general approach, referred to as array comparative genomic hybridization (arrayCGH), used microarrays composed of cloned genomic segments as probes instead of metaphase spreads as the substrate for hybridization (Solinas-Toldo et al. 1997; Pinkel et al. 1998). The advent of in situ synthesis of oligonucleotides on microarray surfaces then allowed the use of oligonucleotide microarrays for high resolution analysis of copy number alterations at a resolution approaching a few hundred to few thousand base pairs (Brennan et al. 2004; Carvalho et al. 2004). Commercial manufacturers such as Agilent, NimbleGen (Roche), Illumina, and Affymetrix are able to print high density oligonucleotide microarrays containing many hundreds of thousands of oligonucleotide probes that query all known non-repetitive portions of the human genome at an average spacing between genomic probes of a few base pairs to hundreds of base pairs, and these arrays are used for analysis of copy number through array-CGH experiments (Barrett et al. 2004; Selzer et al. 2005; Komura et al. 2006; Peiffer et al. 2006). CGH and arrayCGH analyses have been used to find copy number alterations in many human cancers, including prostate cancer (Kallioniemi 2008; Liu et al. 2009). To correct for differences in intensity across the two channels, it is necessary to normalize to the central tendency of the two ratios and therefore, these assays cannot give information as to the overall ploidy of the specimen. For instance, CGH and arrayCGH assays (including SNP arrays as described below) would not be able to detect if a sample showed tetraploidy (four copies of all chromosomes) without any relative gains and losses of individual segments. Rather, these analyses can only provide information regarding the relative gains and losses of chromosomal segments with respect to the central tendency.

2.2.2 Single Nucleotide Polymorphism Microarrays

Single nucleotide polymorphism (SNP) microarrays (Tuefferd et al. 2008; Gunderson 2009) provide the ability to interrogate genotype at thousands to ~one million known SNPs in a parallel fashion. In these arrays, manufacturers such as Illumina and Affymetrix create oligonucleotide microarrays that probe each of up to ~one million SNPs across the human genome. More recently, these arrays also include non-SNP probes that resemble arrayCGH designs (Tuefferd et al. 2008). As an example of a representative assay, Affymetrix's SNP 6.0 microarray platform includes ~one million probe sets targeting SNPs and another ~one million arrayCGH type of probe sets in a single high-density oligonucleotide array. Genomic DNA samples are digested with NspI or StyI restriction enzymes in separate reactions, adapted to universal linkers, pooled together, and subjected to one primer size-restricted PCR designed to optimally amplify genomic segments that are <2 kbp. The products are then further fragmented, labeled and hybridized to the SNP 6.0 microarrays. With such assays, interrogation of genotypes at ~one million SNPs simultaneously has become possible. In addition, identification of genomic alterations such as allele-specific amplifications, deletions, and loss of heterozygosity with normal copy number becomes possible (Liu et al. 2009). In the Affymetrix platform, such analyses are usually carried out with subject-matched tumor and normal specimen pairs. For allele specific copy number analysis, for example, the signal (ratio of normalized intensity in the tumor to that of the normal) at the two probes interrogating each SNP is divided into two bins, a Max allele bin containing the higher of the two signals from the probes interrogating each SNP, and the Min allele bin containing the lower of the two signals from the probes interrogating each SNP. The values in the Max bin are smoothed and the value from the Min bin are smoothed, and plotted to reveal amplifications affecting a single copy in the Max bin "allele" and deletions in the Min bin "allele". Such analyses can indicate if one or both copies of a given locus is/are likely to be amplified or deleted. Additionally, tumors have been known to display normal copy number loss of heterozygosity (analogous to uniparental segmental disomies in genetic disorders). Such regions can be identified by long tracks showing a very low number of heterozygous genotypes greater than can be explained by chance, and that show normal copy number. These regions can arise by deletion of one allele at a given locus, followed by duplication of the other allele at the same locus. Identification of such allele-specific alterations would not be possible by arrayCGH approaches, and represent a major advance of the SNP arrays.

2.2.3 Sequencing Microarrays

A major extension of SNP arrays and arrayCGH microarrays are resequencing arrays (Zheng et al. 2009). With the development of in situ oligonucleotide synthesis at extremely high density, manufacturers such as Affymetrix have begun to offer resequencing arrays with probe sets interrogating single nucleotide sequence changes at every position across a genomic segment. As an example for the design of these arrays, a reference sequence is used to generate every possible ~25mer sequence across a stretch of several thousand base pairs. For each 25mer sequence, four 25mer probesets are synthesized in situ on the surface of the microarray substrate in a high-density format, such that the middle position of each probeset is altered to represent each of the four possible bases. Using such arrays, theoretically, the sequence at each position along the original sequence of interest can be obtained by the degree of hybridization at each of the four probes representing each position.

In order to facilitate high-throughput sample processing, improve performance, and avoid cross hybridization problems that would be encountered with hybridization of unenriched genomic DNA, the assay protocol uses clever methodologies for target amplification by capture and ligation (TACL) to enrich for specific genomic regions of interest, combined with mismatch repair detection (MRD) to separate variant and nonvariant alleles in a nearly homozygous state. These enriched alleles are then "sequenced" by hybridization to the resequencing microarray (Zheng et al. 2009). A major challenge for this type of sequencing is the need for sophisticated bioinformatics algorithms to decipher the signal from the noise produced by cross-hybridization and other issues (Kothiyal et al. 2010). Such issues have limited the ability to hybridize entire human genomic DNA without enrichment of target sequences. Furthermore, this approach is losing momentum with the advent of next generation sequencing (NGS) platforms capable of much higher sequencing capacity. However, advances in bioinformatic approaches for analysis and the ability for high sample throughput still makes this an attractive platform for sequencing many samples in parallel at limited genomic regions (Kothiyal et al. 2010).

2.2.4 Genome-Wide Association Studies

The ability to genotype individuals at hundreds of thousands to millions of SNPs across the genome using SNP microarrays has led to the ability to carry out genome-wide association studies to understand the alleles in the population that confer risk for susceptibility to various diseases including cancer. Over the past several years, research groups have, for the first time, identified risk alleles for common cancers that have been reproduced across several studies and populations (Seng and Seng 2008; Cazier and Tomlinson 2010). Such studies were highlighted by Science Magazine when it hailed the more comprehensive understanding of human genetic variation as the breakthrough of the year in 2007 (Pennisi 2007). A genome wide association study applied to cancer research can involve, for example, a nested case-control design from a large cohort, where germline DNA from all cases (cancer subjects) and controls (matched cancer free subjects) are genotyped at many hundred thousand to a million known SNPs across the human genome using a SNP microarray platform. Using these data, genetic epidemiological and biostatistical analyses are used to identify those alleles that are associated with case/control status, and odds ratios for these alleles are calculated after correcting for multiple hypothesis testing. The identified risk alleles/SNPs are essentially markers indicating that the causal element conferring risk is near the identified risk allele to within a calculatable distance determined by the linkage disequilibrium of the tag SNP with the surrounding SNPs. Using such analyses to study prostate cancer, for instance, several groups have independently confirmed the presence of risk alleles associated with prostate cancer (Guy et al. 2009). Such reproducible identification of risk alleles for prostate cancer, as well as several other cancers such as breast and colorectal cancer, has been facilitated for the first time through these GWAS studies (Easton et al. 2007; Gudmundsson et al. 2007; Yeager et al. 2007; Tenesa et al. 2008; Thomas et al. 2008). Interestingly, in some cases, while any individual SNP has only mild to moderate association, when multiple SNPs are combined, the association becomes significantly higher (Zheng et al. 2008). Nonetheless, one limitation of such studies is that they essentially assume a "common-disease-common-variant" model of predisposition, in which it is assumed that high frequency SNPs (occurring >5% of the population) and low penetrance can have a significant contribution to cancer susceptibility (Cazier and Tomlinson 2010). The approaches detailed above, because they interrogate known common SNPs, are inherently limited in examining the role of rare variants with modest effects in contributing to cancer susceptibility. As the number of known variants increases, and approaches such as next generation sequencing for identification of rare sequence variants in populations are used, the power of these GWAS studies is likely to increase significantly.

2.3 Use of Conventional and Next Generation Sequencing for Genome-Wide Analysis of Genetic Variation/Mutation

Technological breakthroughs in sequencing technologies have been a major driving force for the advancement of molecular biology and molecular genetics in cancer research. The advent of high-throughput Sanger sequencing in the mid- to late-1990s made possible the accelerated completion of the human genome project, which has since revolutionized the pace of discovery in cancer research. Similarly, the advent of next generation sequencing is poised to allow analysis of genomic alterations associated with cancer at an unprecedented scale, and it is anticipated to usher the new era of individualized, rational medicine. Recognizing the tremendous potential of advances in sequencing technologies to revolutionize biomedical research, the X Prize Foundation established the Archon X Prize in Genomics competition (http://genomics.xprize.org/archon-x-prize-for-genomics/prize-overview) to award $10 million to the "first team that can build a device and use it to sequence 100 human genome within 10 days or less, with an accuracy of no more than one error in every 100,000 bases sequenced, with sequences accurately covering at least 98% of the genome, and at a recurring cost of no more than $10,000 per genome." Several companies and teams are now developing technologies to meet this challenge.

2.3.1 High-Throughput Sanger Sequencing

The classical Sanger sequencing method involves the sequencing of a single stranded DNA template by the use of target specific primers, DNA polymerase, and labeled nucleotides and/or chain-termination nucleotides (Sanger and Coulson 1975; Sanger et al. 1977). In the classical version, the sequencing reaction is carried out in four fractions, each containing a full complement of labeled deoxynucleotide triphosphates (dATP, dTTP, dCTP, dGTP), polymerase and primers, but only one of each type of chain termination dideoxynucleotides (ddATP or ddTTP or ddCTP or ddGTP). The resulting reactions produce copies of the original template DNA of varying lengths ranging from a few nucleotides to a few hundred nucleotides (up to 700–1,000 bp). The labeled products of each reaction are then separated on a sequencing gel allowing resolution of sizes separated by a single nucleotide. The DNA bands are then visualized allowing the determination of the positions of each base in the lane corresponding to that base. Major improvements in this classical assay include use of fluorescently labeled chain-terminator nucleotides along with separation by capillary electrophoresis and detection by laser induced fluorescence for automatable high throughput sequencing (Smith et al. 1986). Further improvements to this prototype technology powered the accelerated culmination of the human genome project (Hood and Galas 2003). More recently, this technology was used to sequence ~300,000 amplicons covering all exons of all known genes for cancer and matched normal samples from several individuals for multiple different cancer types (Sjoblom et al. 2006; Wood et al. 2007; Jones et al. 2008; Parsons et al. 2008). The resulting analyses have provided the most complete analysis to date of several human cancers including colon, breast, pancreatic, and brain cancers. However, such large scale analyses are increasingly being facilitated by more cost-effective next generation sequencing methodologies, as these technologies are poised to revolutionize cancer and biomedical research as we head towards the promise of individualized medicine.

2.3.2 Next Generation Sequencing

The term next generation sequencing refers to technologies that have enabled the massively parallel analysis of DNA sequence facilitated through the convergence of advancements in molecular biology, nucleic acid chemistry and biochemistry, computational biology, and electrical and mechanical engineering. The current next generation sequencing technologies are capable of sequencing tens to hundreds of millions of DNA templates simultaneously and generate >4 Gigabases of sequence in a single day. These technologies have largely started to replace high throughput Sanger sequencing for large-scale genomic projects, and have created significant enthusiasm for the advent of a new era of individualized medicine.

2.3.3 Overview of Commercialized Next Generation Sequencing Platforms

Given the promise of and demand for next generation sequencing technologies, there has been intense competition from companies for development of NGS platforms. 454 life technologies, later acquired by Roche, was the first to release an NGS platform. Solexa, now part of Illumina, released the next platform, with Applied Biosystems marketing the third commercialized platform which it acquired from Agencourt. Helicos was the first company to release a single molecule sequencing NGS platform, and more recently several new companies have entered the arena, including Complete Genomics, Pacific Biosciences, and Ion Torrents, with more to follow in the near future. A more detailed discussion of the technologies from some of the currently commercialized platforms is given in Chap. 6. Here, we can focus on the broad innovations used to facilitate discovery of genomic variation using these platforms.

The major components of the NGS workflow that are generically applicable to all of the current technologies are library choice/construction, preparation of libraries for sequencing, and massively parallel sequencing. We will discuss each of these components below and highlight the broad similarities and differences between platforms along with the strengths and weaknesses for analysis of sequence variation.

2.3.3.1 Library Choice and Construction

Two major types of libraries are used depending on the application: fragment library, and mate-paired libraries. In a fragment library, genomic DNA from a sample is randomly fragmented to a small modal size, typically just 1–5 times the size of the sequencing platform's read length. Sequencing adaptors are then attached to these library molecules to facilitate sequencing from a single end of each DNA fragment in the library. More recently, it has become possible to sequence from both ends of such library DNA fragments in a process referred to as fragment paired-end sequencing. Fragment libraries are extremely useful for analysis of single nucleotide substitutions/variations. Each fragment in the library produces a single read and multiple overlapping fragments are sequenced for each position in the genome. A coverage of >30× is usually needed to confidently decipher true variation from sequencing errors and for robustly distinguishing homozygous and heterozygous SNPs. Additionally, fragment libraries can also provide some information on genomic copy number. This can be done by taking all of the fragment library reads within fixed genomic bins and carrying out analyses to assess whether the number of reads observed is different than the number expected by random chance (e.g. Xie and Tammi 2009). Such methods are an extension of digital karyotyping analyses (Wang et al. 2002). Fragment libraries can also be target enriched with microarray or solution-based hybrid capture strategies for targeted resequencing projects (Albert et al. 2007; Gnirke et al. 2009). In these analyses, a fragment library is first prepared. Next, the library is subjected to target sequence enrichment by hybridization to oligonucleotides complementary to desired targets. The oligonucleotide "baits" can be immobilized on the surface of a microarray that is very similar to the microarrays described in the previous section. Agilent and Nimblegen among other companies have begun to offer this as a standard or custom design product. More recently, the oligonucleotides are synthesized in situ on microarrays, then released by cleavage from the microarray, amplified, and modified with biotin and immobilized on magnetic beads to allow solution-based capture of target sequences (Gnirke et al. 2009). Agilent markets this as their SureSelect solution-capture-based target enrichment strategy, and kits have been released for use with the Illumina and SOLiD NGS platforms. Such approaches have allowed targeted resequencing of any desired portion of the genome, such as all exons in the human genome (Maher 2009).

A mate-paired library is constructed by first randomly shearing or fragmenting genomic DNA to a modal size that is typically >1,000 bps, which is in significant excess of the read lengths produced by most of the currently commercialized platforms. This library is then size-separated on a gel, and the portion of the library corresponding to a specific size range, e.g. 2–3 kbp, is excised and purified. These fragments are then circularized via ligation of an adapter under conditions that promote circularization with the adaptor. This geometry allows generation of a library consisting of DNA fragments comprised of sub-fragments from the two ends of the original size-selected DNA library juxtaposed to each other. The two mate-paired sub-fragments are then sequenced to reveal the sequences underlying the two ends of each 2–3 kbp library template. Because we know a priori the possible distances between the two sequences comprising the mate-paired read, after alignment to the reference genome, we can calculate whether there was likely to be an amplification, deletion, or translocation between the mate-paired sequences. Similarly, the orientation of the sequences can be used to detect inversions. Therefore, mate-paired libraries not only provide information on single nucleotide substitutions, but also on structural variation in the genome, as has been demonstrated in several recent reports (Korbel et al. 2007; McKernan et al. 2009).

With the advent of more recent NGS platforms, other library types are also possible. Pacific Biosciences has developed ultra-long read lengths >1,000 base pairs and have deployed these highly processive reads to generate repeated serial reads of both strands of double strand DNA after circularization of a fragment library with hairpin adaptors. The resulting "SMRT Bell" libraries allow high fidelity sequencing where the accuracy increases with the number of times the polymerase traverses the circular SMRT Bell fragments (http://www.pacificbiosciences.com/). This company is also developing strobe-sequencing, where the progress of the processive polymerase in copying long template DNA is recorded in an on-off periodic fashion as a way to generate several mate-tags of sequence from a long DNA template, with all tags oriented in the same direction. Other companies such as Complete Genomics have introduced highly complex library generation strategies involving serial cutting and circularization to fabricate DNA nanoballs for unchained ligation based sequencing (Drmanac et al. 2010). This strategy has been used for sequencing of whole human genomes for identification of single nucleotide variation (Roach et al. 2010). Other library configurations and geometries are likely to surface as the diversity of NGS platforms increases.

2.3.3.2 Preparation of Libraries for Sequencing on NGS Platforms

The steps involved in preparing libraries for sequencing on a specific NGS platform are usually tailor made for that platform. For the Roche 454 and Applied Biosystems SOLiD systems, this involves emulsion PCR (Dressman et al. 2003) to amplify each individual template DNA molecule clonally on the surface of a bead. In emulsion PCR, individual DNA templates are sequestered along with PCR reagents and a primer coated bead within an aqueous droplet surrounded by a hydrophobic shell within an oil-in-water emulsion. Subjecting these droplets to PCR allows clonal amplification of each template DNA molecule onto the surface of the bead. In the case of Roche 454, the beads are then deposited in pico-litre scale wells of a plate which serve as the substrate for sequencing on the instrument (Margulies et al. 2005). In the case of Applied Biosystems, the clonally amplified DNA molecules on the surface of the bead are end-modified and covalently and randomly attached to the surface of a glass slide, which is then loaded for sequencing on the instrument. Recent improvements in the automation of the emulsion PCR process have made these steps more streamlined. For the Illumina/Solexa Genome Analyzer and HiSeq platforms, DNA libraries are subjected to clonal bridge amplification and cluster generation in situ on the surface of lanes in a flow cell. These flow cells are then subjected to sequencing analysis. For Helicos, library generation is somewhat simpler, and does not require any major amplification steps. In their true single molecule sequencing (tSMS) platform, library fragments are tailed with poly-adenosine and hybridized onto oligo-dT primer-conjugated flow cells, which are then subjected to sequencing via extension from the oligo-dT primers (Harris et al. 2008).

2.3.3.3 Massively Parallel Sequencing of Libraries on NGS Instruments

Each of the currently commercialized NGS platforms uses a distinct set of chemistries to allow massively parallel sequencing of many millions to billions of template DNA molecules. The differences in chemistries confer various strengths and weaknesses to each of the platforms. Because these technologies are rapidly changing, we will focus our discussion on the broad characteristics of the chemistries that are likely to remain stable for the currently commercialized platforms and only touch briefly on up-and-coming platforms that have not yet seen widespread adoption.

The Roche 454 system (http://www.454.com/) uses a sequence-by-synthesis strategy in which DNA templates on the surface of a bead are copied by a DNA polymerase which is forced to add only one nucleotide species at a time by cycling the flow of each nucleotide in turn and repeating these cycles for several iterations (Margulies et al. 2005). The pyrophosphates released by the polymerase are converted to light by a pyrosequencing process where the amount of light emitted can be used to calculate the number of a specific nucleotide added at each cycle. One somewhat persistent problem with this method is that mononucleotide repeat tracks (e.g. a run of 12 adenines in a row) can give rise to errors. This method allows sequencing read lengths of >400 bps in current implementations. However, the overall throughput is somewhat limited by the number of picoliter wells on a plate that can be sequenced, and this platform currently has the lowest sequence capacity per time or per dollar compared to the other commercialized platforms.

Illumina (http://www.illumina.com/) and Helicos (http://www.helicosbio.com/) also use a sequence-by-synthesis strategy, but avoid the errors associated with mono-nucleotide runs by using fluorescently labeled reversible chain terminator nucleotides allowing controlled addition of only a single nucleotide at a time. Because these platforms halt at the addition of every single nucleotide, the coupling efficiencies become limiting and read lengths on these platforms are typically less than 100 bp. In the case of Helicos, which uses tSMS technology, there appears to be a persistent issue of dark bases in which the addition of a nucleotide is not associated with fluorescence generation. This will probably be an issue with other emerging tSMS platforms.

The Applied Biosystems (http://solid.appliedbiosystems.com) (now Life Technologies) SOLiD platform uses a sequence-by-ligation approach in which a DNA ligase, instead of a DNA polymerase, is used to assess sequence via sequential ligation of fluorescently labeled oligonucleotide probes that can interrogate every combination of two adjacent bases (16 combinations possible). However, there are only four different fluorescent dyes, with each dye interrogating one of four possible dinucleotide combinations. Because of this, an individual ligation reaction does not uniquely identify a given base or dinucleotide combination. Each base in the sequence is interrogated twice in this degenerate fashion and the combined data across an entire read can be deconvoluted to decipher the actual sequence. The first step of the sequencing reaction is to anneal a sequencing primer to the P1 adaptor on the library template and then to add a mixture of the 16 possible labeled probes. The appropriate di-base probe binds to the first and second base of the template and is ligated to the sequencing primer. The fluorophore associated with this probe is then registered and the probe is enzymatically processed to allow sequential ligation of another probe to interrogate the sixth and seventh bases. This process is repeated a total of eight additional times for a total of ten ligation steps for the first primer. After the last ligation step, the reaction is "reset" by denaturing and washing away the newly synthesized strand from the template DNA that is covalently linked to the bead (see emulsion PCR description above). A new sequencing primer that is complementary to a sequence that is off-set by one base from the first primer is then annealed so that the first ligation reaction stemming from this sequencing primer interrogates the last base of the adaptor sequence (position 0) and the first base of the template. This primer also goes through ten ligation steps. There are a total of five different sequencing primers that each undergo a total of ten ligation steps. This results in each base being interrogated twice and a sequencing length of 50 base pairs.

2.3.4 The Near and Long Term Horizon

Each of the platforms described above are routinely making advancements in sequencing throughput both in terms of time and cost per Gbp of sequence output.

In the meanwhile, other platforms such as Complete Genomics (http://www.completegenomics.com/), Pacific Biosciences (http://www.pacificbiosciences.com/), Ion Torrent (http://www.iontorrent.com/), and possibly several other players are preparing to enter the market with systems that can be deployed to individual labs and genome centers. As a result of this intense competition, cost and time per Gbp of sequence produced is rapidly declining. In the near future, it will be possible to carry out large-scale genome-wide association studies to identify even rare variants that are causally linked with disease. Additionally, it may become routine to sequence the entire genome of cancer and normal specimens from individuals with cancer to obtain a comprehensive list of all mutations and pathways. This information can serve as a source of individualized biomarkers to track disease burden in these individuals as has recently been shown (Leary et al. 2010). In addition, such information can provide individualized guidance for therapeutic decision making. The key will be to develop, in parallel, the computational, statistical and informatics solutions to harness the power of these increasingly cost-effective technologies and deploy them at the population scale.

References

Albert TJ, Molla MN, Muzny DM, Nazareth L, Wheeler D, Song X, Richmond TA, Middle CM, Rodesch MJ, Packard CJ, Weinstock GM, Gibbs RA (2007) Direct selection of human genomic loci by microarray hybridization. Nat Methods 4:903–905.

Barrett MT, Scheffer A, Ben-Dor A, Sampas N, Lipson D, Kincaid R, Tsang P, Curry B, Baird K, Meltzer PS, Yakhini Z, Bruhn L, Laderman S (2004) Comparative genomic hybridization using oligonucleotide microarrays and total genomic DNA. Proc Natl Acad Sci U S A 101:17765–17770.

Bauman JG, Wiegant J, Borst P, van Duijn P (1980) A new method for fluorescence microscopical localization of specific DNA sequences by in situ hybridization of fluorochromelabelled RNA. Exp Cell Res 128:485–490.

Braig M, Schmitt CA (2006) Oncogene-induced senescence: putting the brakes on tumor development. Cancer Res 66:2881–2884.

Brennan C, Zhang Y, Leo C, Feng B, Cauwels C, Aguirre AJ, Kim M, Protopopov A, Chin L (2004) High-resolution global profiling of genomic alterations with long oligonucleotide microarray. Cancer Res 64:4744–4748.

Carvalho B, Ouwerkerk E, Meijer GA, Ylstra B (2004) High resolution microarray comparative genomic hybridisation analysis using spotted oligonucleotides. J Clin Pathol 57:644–646.

Cazier JB, Tomlinson I (2010) General lessons from large-scale studies to identify human cancer predisposition genes. J Pathol 220:255–262.

Cremer T, Lichter P, Borden J, Ward DC, Manuelidis L (1988) Detection of chromosome aberrations in metaphase and interphase tumor cells by in situ hybridization using chromosome-specific library probes. Hum Genet 80:235–246.

Dressman D, Yan H, Traverso G, Kinzler KW, Vogelstein B (2003) Transforming single DNA molecules into fluorescent magnetic particles for detection and enumeration of genetic variations. Proc Natl Acad Sci U S A 100:8817–8822.

Drets ME, Shaw MW (1971) Specific banding patterns of human chromosomes. Proc Natl Acad Sci U S A 68:2073–2077.

Drmanac R, Sparks AB, Callow MJ, Halpern AL, Burns NL, Kermani BG, Carnevali P, Nazarenko I, Nilsen GB, Yeung G, Dahl F, Fernandez A, Staker B, Pant KP, Baccash J, Borcherding AP,

Brownley A, Cedeno R, Chen L, Chernikoff D, Cheung A, Chirita R, Curson B, Ebert JC, Hacker CR, Hartlage R, Hauser B, Huang S, Jiang Y, Karpinchyk V, Koenig M, Kong C, Landers T, Le C, Liu J, McBride CE, Morenzoni M, Morey RE, Mutch K, Perazich H, Perry K, Peters BA, Peterson J, Pethiyagoda CL, Pothuraju K, Richter C, Rosenbaum AM, Roy S, Shafto J, Sharanhovich U, Shannon KW, Sheppy CG, Sun M, Thakuria JV, Tran A, Vu D, Zaranek AW, Wu X, Drmanac S, Oliphant AR, Banyai WC, Martin B, Ballinger DG, Church GM, Reid CA (2010) Human genome sequencing using unchained base reads on self-assembling DNA nanoarrays. Science 327:78–81.

du Manoir S, Speicher MR, Joos S, Schrock E, Popp S, Dohner H, Kovacs G, Robert-Nicoud M, Lichter P, Cremer T (1993) Detection of complete and partial chromosome gains and losses by comparative genomic in situ hybridization. Hum Genet 90:590–610.

Easton DF, Pooley KA, Dunning AM, Pharoah PD, Thompson D, Ballinger DG, Struewing JP, Morrison J, Field H, Luben R, Wareham N, Ahmed S, Healey CS, Bowman R, Meyer KB, Haiman CA, Kolonel LK, Henderson BE, Le Marchand L, Brennan P, Sangrajrang S, Gaborieau V, Odefrey F, Shen CY, Wu PE, Wang HC, Eccles D, Evans DG, Peto J, Fletcher O, Johnson N, Seal S, Stratton MR, Rahman N, Chenevix-Trench G, Bojesen SE, Nordestgaard BG, Axelsson CK, Garcia-Closas M, Brinton L, Chanock S, Lissowska J, Peplonska B, Nevanlinna H, Fagerholm R, Eerola H, Kang D, Yoo KY, Noh DY, Ahn SH, Hunter DJ, Hankinson SE, Cox DG, Hall P, Wedren S, Liu J, Low YL, Bogdanova N, Schurmann P, Dork T, Tollenaar RA, Jacobi CE, Devilee P, Klijn JG, Sigurdson AJ, Doody MM, Alexander BH, Zhang J, Cox A, Brock IW, MacPherson G, Reed MW, Couch FJ, Goode EL, Olson JE, Meijers-Heijboer H, van den Ouweland A, Uitterlinden A, Rivadeneira F, Milne RL, Ribas G, Gonzalez-Neira A, Benitez J, Hopper JL, McCredie M, Southey M, Giles GG, Schroen C, Justenhoven C, Brauch H, Hamann U, Ko YD, Spurdle AB, Beesley J, Chen X, Mannermaa A, Kosma VM, Kataja V, Hartikainen J, Day NE, Cox DR, Ponder BA (2007) Genome-wide association study identifies novel breast cancer susceptibility loci. Nature 447:1087–1093.

Gnirke A, Melnikov A, Maguire J, Rogov P, LeProust EM, Brockman W, Fennell T, Giannoukos G, Fisher S, Russ C, Gabriel S, Jaffe DB, Lander ES, Nusbaum C (2009) Solution hybrid selection with ultra-long oligonucleotides for massively parallel targeted sequencing. Nat Biotechnol 27:182–189.

Gudmundsson J, Sulem P, Manolescu A, Amundadottir LT, Gudbjartsson D, Helgason A, Rafnar T, Bergthorsson JT, Agnarsson BA, Baker A, Sigurdsson A, Benediktsdottir KR, Jakobsdottir M, Xu J, Blondal T, Kostic J, Sun J, Ghosh S, Stacey SN, Mouy M, Saemundsdottir J, Backman VM, Kristjansson K, Tres A, Partin AW, Albers-Akkers MT, Godino-Ivan Marcos J, Walsh PC, Swinkels DW, Navarrete S, Isaacs SD, Aben KK, Graif T, Cashy J, Ruiz-Echarri M, Wiley KE, Suarez BK, Witjes JA, Frigge M, Ober C, Jonsson E, Einarsson GV, Mayordomo JI, Kiemeney LA, Isaacs WB, Catalona WJ, Barkardottir RB, Gulcher JR, Thorsteinsdottir U, Kong A, Stefansson K (2007) Genome-wide association study identifies a second prostate cancer susceptibility variant at 8q24. Nat Genet 39:631–637.

Gunderson KL (2009) Whole-genome genotyping on bead arrays. Methods Mol Biol 529:197–213.

Guy M, Kote-Jarai Z, Giles GG, Al Olama AA, Jugurnauth SK, Mulholland S, Leongamornlert DA, Edwards SM, Morrison J, Field HI, Southey MC, Severi G, Donovan JL, Hamdy FC, Dearnaley DP, Muir KR, Smith C, Bagnato M, Ardern-Jones AT, Hall AL, O'Brien LT, Gehr-Swain BN, Wilkinson RA, Cox A, Lewis S, Brown PM, Jhavar SG, Tymrakiewicz M, Lophatananon A, Bryant SL, Horwich A, Huddart RA, Khoo VS, Parker CC, Woodhouse CJ, Thompson A, Christmas T, Ogden C, Fisher C, Jameson C, Cooper CS, English DR, Hopper JL, Neal DE, Easton DF, Eeles RA (2009) Identification of new genetic risk factors for prostate cancer. Asian J Androl 11:49–55.

Haenszel W, Kurihara M (1968) Studies of Japanese migrants. I. Mortality from cancer and other diseases among Japanese in the United States. J Natl Cancer Inst 40:43–68.

Hanahan D, Weinberg RA (2000) The hallmarks of cancer. Cell 100:57–70.

Harris TD, Buzby PR, Babcock H, Beer E, Bowers J, Braslavsky I, Causey M, Colonell J, Dimeo J, Efcavitch JW, Giladi E, Gill J, Healy J, Jarosz M, Lapen D, Moulton K, Quake SR, Steinmann K, Thayer E, Tyurina A, Ward R, Weiss H, Xie Z (2008) Single-molecule DNA sequencing of a viral genome. Science 320:106–109.

Hood L, Galas D (2003) The digital code of DNA. Nature 421:444–448.

Houlston RS, Peto J (1996) In: Eeles RA, Ponder BAJ, Easton DF, Horwich A (eds) Genetic predisposition to cancer. Chapman & Hall, London, pp 208–226.

Hsing AW, Tsao L, Devesa SS (2000) International trends and patterns of prostate cancer incidence and mortality. Int J Cancer 85:60–67.

Jones PA, Laird PW (1999) Cancer epigenetics comes of age. Nat Genet 21:163–167.

Jones S, Zhang X, Parsons DW, Lin JC, Leary RJ, Angenendt P, Mankoo P, Carter H, Kamiyama H, Jimeno A, Hong SM, Fu B, Lin MT, Calhoun ES, Kamiyama M, Walter K, Nikolskaya T, Nikolsky Y, Hartigan J, Smith DR, Hidalgo M, Leach SD, Klein AP, Jaffee EM, Goggins M, Maitra A, Iacobuzio-Donahue C, Eshleman JR, Kern SE, Hruban RH, Karchin R, Papadopoulos N, Parmigiani G, Vogelstein B, Velculescu VE, Kinzler KW (2008) Core signaling pathways in human pancreatic cancers revealed by global genomic analyses. Science 321:1801–1806.

Kallioniemi A (2008) CGH microarrays and cancer. Curr Opin Biotechnol 19:36–40.

Kallioniemi A, Kallioniemi OP, Sudar D, Rutovitz D, Gray JW, Waldman F, Pinkel D (1992) Comparative genomic hybridization for molecular cytogenetic analysis of solid tumors. Science 258:818–821.

Kinzler KW, Vogelstein B (1996) Lessons from hereditary colorectal cancer. Cell 87:159–170.

Knudson AG Jr (1971) Mutation and cancer: statistical study of retinoblastoma. Proc Natl Acad Sci U S A 68:820–823.

Knudson AG (2001) Two genetic hits (more or less) to cancer. Nat Rev Cancer 1:157–162.

Komura D, Shen F, Ishikawa S, Fitch KR, Chen W, Zhang J, Liu G, Ihara S, Nakamura H, Hurles ME, Lee C, Scherer SW, Jones KW, Shapero MH, Huang J, Aburatani H (2006) Genome-wide detection of human copy number variations using high-density DNA oligonucleotide arrays. Genome Res 16:1575–1584.

Korbel JO, Urban AE, Affourtit JP, Godwin B, Grubert F, Simons JF, Kim PM, Palejev D, Carriero NJ, Du L, Taillon BE, Chen Z, Tanzer A, Saunders AC, Chi J, Yang F, Carter NP, Hurles ME, Weissman SM, Harkins TT, Gerstein MB, Egholm M, Snyder M (2007) Paired-end mapping reveals extensive structural variation in the human genome. Science 318:420–426.

Kothiyal P, Cox S, Ebert J, Husami A, Kenna MA, Greinwald JH, Aronow BJ, Rehm HL (2010) High-throughput detection of mutations responsible for childhood hearing loss using resequencing microarrays. BMC Biotechnol 10:10.

Leary RJ, Kinde I, Diehl F, Schmidt K, Clouser C, Duncan C, Antipova A, Lee C, McKernan K, De La Vega FM, Kinzler KW, Vogelstein B, Diaz LA Jr, Velculescu VE (2010) Development of personalized tumor biomarkers using massively parallel sequencing. Sci Transl Med 2:20–14.

Lichtenstein P, Holm NV, Verkasalo PK, Iliadou A, Kaprio J, Koskenvuo M, Pukkala E, Skytthe A, Hemminki K (2000) Environmental and heritable factors in the causation of cancer–analyses of cohorts of twins from Sweden, Denmark, and Finland. N Engl J Med 343:78–85.

Lichter P, Cremer T, Borden J, Manuelidis L, Ward DC (1988) Delineation of individual human chromosomes in metaphase and interphase cells by in situ suppression hybridization using recombinant DNA libraries. Hum Genet 80:224–234.

Liu W, Laitinen S, Khan S, Vihinen M, Kowalski J, Yu G, Chen L, Ewing CM, Eisenberger MA, Carducci MA, Nelson WG, Yegnasubramanian S, Luo J, Wang Y, Xu J, Isaacs WB, Visakorpi T, Bova GS (2009) Copy number analysis indicates monoclonal origin of lethal metastatic prostate cancer. Nat Med 15:559–565.

Maher B (2009) Exome sequencing takes centre stage in cancer profiling. Nature 459:146–147.

Margulies M, Egholm M, Altman WE, Attiya S, Bader JS, Bemben LA, Berka J, Braverman MS, Chen YJ, Chen Z, Dewell SB, Du L, Fierro JM, Gomes XV, Godwin BC, He W, Helgesen S, Ho CH, Irzyk GP, Jando SC, Alenquer ML, Jarvie TP, Jirage KB, Kim JB, Knight JR, Lanza JR, Leamon JH, Lefkowitz SM, Lei M, Li J, Lohman KL, Lu H, Makhijani VB, McDade KE,

McKenna MP, Myers EW, Nickerson E, Nobile JR, Plant R, Puc BP, Ronan MT, Roth GT, Sarkis GJ, Simons JF, Simpson JW, Srinivasan M, Tartaro KR, Tomasz A, Vogt KA, Volkmer GA, Wang SH, Wang Y, Weiner MP, Yu P, Begley RF, Rothberg JM (2005) Genome sequencing in microfabricated high-density picolitre reactors. Nature 437:376–380.

McKernan KJ, Peckham HE, Costa GL, McLaughlin SF, Fu Y, Tsung EF, Clouser CR, Duncan C, Ichikawa JK, Lee CC, Zhang Z, Ranade SS, Dimalanta ET, Hyland FC, Sokolsky TD, Zhang L, Sheridan A, Fu H, Hendrickson CL, Li B, Kotler L, Stuart JR, Malek JA, Manning JM, Antipova AA, Perez DS, Moore MP, Hayashibara KC, Lyons MR, Beaudoin RE, Coleman BE, Laptewicz MW, Sannicandro AE, Rhodes MD, Gottimukkala RK, Yang S, Bafna V, Bashir A, MacBride A, Alkan C, Kidd JM, Eichler EE, Reese MG, De La Vega FM, Blanchard AP (2009) Sequence and structural variation in a human genome uncovered by short-read, massively parallel ligation sequencing using two-base encoding. Genome Res 19:1527–1541.

Parsons DW, Jones S, Zhang X, Lin JC, Leary RJ, Angenendt P, Mankoo P, Carter H, Siu IM, Gallia GL, Olivi A, McLendon R, Rasheed BA, Keir S, Nikolskaya T, Nikolsky Y, Busam DA, Tekleab H, Diaz LA Jr, Hartigan J, Smith DR, Strausberg RL, Marie SK, Shinjo SM, Yan H, Riggins GJ, Bigner DD, Karchin R, Papadopoulos N, Parmigiani G, Vogelstein B, Velculescu VE, Kinzler KW (2008) An integrated genomic analysis of human glioblastoma multiforme. Science 321:1807–1812.

Peiffer DA, Le JM, Steemers FJ, Chang W, Jenniges T, Garcia F, Haden K, Li J, Shaw CA, Belmont J, Cheung SW, Shen RM, Barker DL, Gunderson KL (2006) High-resolution genomic profiling of chromosomal aberrations using Infinium whole-genome genotyping. Genome Res 16:1136–1148.

Pennisi E (2007) Breakthrough of the year. Human genetic variation. Science 318:1842–1843.

Pinkel D, Landegent J, Collins C, Fuscoe J, Segraves R, Lucas J, Gray J (1988) Fluorescence in situ hybridization with human chromosome-specific libraries: detection of trisomy 21 and translocations of chromosome 4. Proc Natl Acad Sci U S A 85:9138–9142.

Pinkel D, Segraves R, Sudar D, Clark S, Poole I, Kowbel D, Collins C, Kuo WL, Chen C, Zhai Y, Dairkee SH, Ljung BM, Gray JW, Albertson DG (1998) High resolution analysis of DNA copy number variation using comparative genomic hybridization to microarrays. Nat Genet 20:207–211.

Roach JC, Glusman G, Smit AF, Huff CD, Hubley R, Shannon PT, Rowen L, Pant KP, Goodman N, Bamshad M, Shendure J, Drmanac R, Jorde LB, Hood L, Galas DJ (2010) Analysis of genetic inheritance in a family quartet by whole-genome sequencing. Science 328:636–639.

Rowley JD (1973) Letter: A new consistent chromosomal abnormality in chronic myelogenous leukaemia identified by quinacrine fluorescence and Giemsa staining. Nature 243:290–293.

Sanger F, Coulson AR (1975) A rapid method for determining sequences in DNA by primed synthesis with DNA polymerase. J Mol Biol 94:441–448.

Sanger F, Nicklen S, Coulson AR (1977) DNA sequencing with chain-terminating inhibitors. Proc Natl Acad Sci U S A 74:5463–5467.

Schrock E, du Manoir S, Veldman T, Schoell B, Wienberg J, Ferguson-Smith MA, Ning Y, Ledbetter DH, Bar-Am I, Soenksen D, Garini Y, Ried T (1996) Multicolor spectral karyotyping of human chromosomes. Science 273:494–497.

Selzer RR, Richmond TA, Pofahl NJ, Green RD, Eis PS, Nair P, Brothman AR, Stallings RL (2005) Analysis of chromosome breakpoints in neuroblastoma at sub-kilobase resolution using fine-tiling oligonucleotide array CGH. Genes Chromosomes Cancer 44:305–319.

Seng KC, Seng CK (2008) The success of the genome-wide association approach: a brief story of a long struggle. Eur J Hum Genet 16:554–564.

Shimizu H, Ross RK, Bernstein L, Yatani R, Henderson BE, Mack TM (1991) Cancers of the prostate and breast among Japanese and white immigrants in Los Angeles County. Br J Cancer 63:963–966.

Sjoblom T, Jones S, Wood LD, Parsons DW, Lin J, Barber TD, Mandelker D, Leary RJ, Ptak J, Silliman N, Szabo S, Buckhaults P, Farrell C, Meeh P, Markowitz SD, Willis J, Dawson D, Willson JK, Gazdar AF, Hartigan J, Wu L, Liu C, Parmigiani G, Park BH, Bachman KE, Papadopoulos N, Vogelstein B, Kinzler KW, Velculescu VE (2006) The consensus coding sequences of human breast and colorectal cancers. Science 314:268–274.

Smith LM, Sanders JZ, Kaiser RJ, Hughes P, Dodd C, Connell CR, Heiner C, Kent SB, Hood LE (1986) Fluorescence detection in automated DNA sequence analysis. Nature 321:674–679.
Solinas-Toldo S, Lampel S, Stilgenbauer S, Nickolenko J, Benner A, Dohner H, Cremer T, Lichter P (1997) Matrix-based comparative genomic hybridization: biochips to screen for genomic imbalances. Genes Chromosomes Cancer 20:399–407.
Tenesa A, Farrington SM, Prendergast JG, Porteous ME, Walker M, Haq N, Barnetson RA, Theodoratou E, Cetnarskyj R, Cartwright N, Semple C, Clark AJ, Reid FJ, Smith LA, Kavoussanakis K, Koessler T, Pharoah PD, Buch S, Schafmayer C, Tepel J, Schreiber S, Volzke H, Schmidt CO, Hampe J, Chang-Claude J, Hoffmeister M, Brenner H, Wilkening S, Canzian F, Capella G, Moreno V, Deary IJ, Starr JM, Tomlinson IP, Kemp Z, Howarth K, Carvajal-Carmona L, Webb E, Broderick P, Vijayakrishnan J, Houlston RS, Rennert G, Ballinger D, Rozek L, Gruber SB, Matsuda K, Kidokoro T, Nakamura Y, Zanke BW, Greenwood CM, Rangrej J, Kustra R, Montpetit A, Hudson TJ, Gallinger S, Campbell H, Dunlop MG (2008) Genome-wide association scan identifies a colorectal cancer susceptibility locus on 11q23 and replicates risk loci at 8q24 and 18q21. Nat Genet 40:631–637.
Thomas G, Jacobs KB, Yeager M, Kraft P, Wacholder S, Orr N, Yu K, Chatterjee N, Welch R, Hutchinson A, Crenshaw A, Cancel-Tassin G, Staats BJ, Wang Z, Gonzalez-Bosquet J, Fang J, Deng X, Berndt SI, Calle EE, Feigelson HS, Thun MJ, Rodriguez C, Albanes D, Virtamo J, Weinstein S, Schumacher FR, Giovannucci E, Willett WC, Cussenot O, Valeri A, Andriole GL, Crawford ED, Tucker M, Gerhard DS, Fraumeni JF Jr, Hoover R, Hayes RB, Hunter DJ, Chanock SJ (2008) Multiple loci identified in a genome-wide association study of prostate cancer. Nat Genet 40:310–315.
Tuefferd M, De Bondt A, Van Den Wyngaert I, Talloen W, Verbeke T, Carvalho B, Clevert DA, Alifano M, Raghavan N, Amaratunga D, Gohlmann H, Broet P, Camilleri-Broet S (2008) Genome-wide copy number alterations detection in fresh frozen and matched FFPE samples using SNP 6.0 arrays. Genes Chromosomes Cancer 47:957–964.
Wang TL, Maierhofer C, Speicher MR, Lengauer C, Vogelstein B, Kinzler KW, Velculescu VE (2002) Digital karyotyping. Proc Natl Acad Sci U S A 99:16156–16161.
Whittemore AS, Kolonel LN, Wu AH, John EM, Gallagher RP, Howe GR, Burch JD, Hankin J, Dreon DM, West DW et al (1995) Prostate cancer in relation to diet, physical activity, and body size in blacks, whites, and Asians in the United States and Canada. J Natl Cancer Inst 87:652–661.
Wood LD, Parsons DW, Jones S, Lin J, Sjoblom T, Leary RJ, Shen D, Boca SM, Barber T, Ptak J, Silliman N, Szabo S, Dezso Z, Ustyanksky V, Nikolskaya T, Nikolsky Y, Karchin R, Wilson PA, Kaminker JS, Zhang Z, Croshaw R, Willis J, Dawson D, Shipitsin M, Willson JK, Sukumar S, Polyak K, Park BH, Pethiyagoda CL, Pant PV, Ballinger DG, Sparks AB, Hartigan J, Smith DR, Suh E, Papadopoulos N, Buckhaults P, Markowitz SD, Parmigiani G, Kinzler KW, Velculescu VE, Vogelstein B (2007) The genomic landscapes of human breast and colorectal cancers. Science 318:1108–1113.
Xie C, Tammi MT (2009) CNV-seq, a new method to detect copy number variation using high-throughput sequencing. BMC Bioinformatics 10:80.
Yeager M, Orr N, Hayes RB, Jacobs KB, Kraft P, Wacholder S, Minichiello MJ, Fearnhead P, Yu K, Chatterjee N, Wang Z, Welch R, Staats BJ, Calle EE, Feigelson HS, Thun MJ, Rodriguez C, Albanes D, Virtamo J, Weinstein S, Schumacher FR, Giovannucci E, Willett WC, Cancel-Tassin G, Cussenot O, Valeri A, Andriole GL, Gelmann EP, Tucker M, Gerhard DS, Fraumeni JF Jr, Hoover R, Hunter DJ, Chanock SJ, Thomas G (2007) Genome-wide association study of prostate cancer identifies a second risk locus at 8q24. Nat Genet 39:645–649.
Zheng SL, Sun J, Wiklund F, Smith S, Stattin P, Li G, Adami HO, Hsu FC, Zhu Y, Balter K, Kader AK, Turner AR, Liu W, Bleecker ER, Meyers DA, Duggan D, Carpten JD, Chang BL, Isaacs WB, Xu J, Gronberg H (2008) Cumulative association of five genetic variants with prostate cancer. N Engl J Med 358:910–919.
Zheng J, Moorhead M, Weng L, Siddiqui F, Carlton VE, Ireland JS, Lee L, Peterson J, Wilkins J, Lin S, Kan Z, Seshagiri S, Davis RW, Faham M (2009) High-throughput, high-accuracy array-based resequencing. Proc Natl Acad Sci U S A 106:6712–6717.

Chapter 3
Examining DNA–Protein Interactions with Genome-Wide Chromatin Immunoprecipitation Analysis

Esteban Ballestar and Manel Esteller

Abstract Understanding the mechanisms by which genomic information is hierarchically organized and used by different cell and tissue types under different physiological conditions requires the detailed analysis of the chromatin structure and nuclear factor distribution throughout the entire genome. Chromatin organization and histone modification patterns ultimately define cell identity, and the occurrence of aberrant changes in chromatin result in malfunction and disease. The strategy of systematically mapping the distribution of histone modifications, nucleosome positioning, and nuclear factor occupancy requires key methods for mapping DNA-protein interactions at the genome-wide level.

The use of chromatin immunoprecipitation (ChIP) assays, where an immunoprecipitating antibody against a particular factor or histone modification is used to enrich chromatin fractions in the sequences to which the protein is bound, has been very useful for defining the histone modification status and nuclear factor association at specific sites. More recently, the combination of ChIP assays with hybridization of microarrays (ChIP-chip) or with next generation massively parallel sequencing (ChIP-seq) has become an important strategy for the acquisition of this type of information at the genome-wide level. In this chapter, we present a critical overview of the considerations necessary for the design and execution of successful genome-wide ChIP experiments.

3.1 Introduction

The functionality of the genome in the context of a eukaryotic cell is defined by patterns of chromatin structure throughout the sequence, i.e., patterns of histone modifications and different nuclear factors associated specifically with their

E. Ballestar (✉)
Cancer Epigenetics Group, Spanish National Cancer Research Centre (CNIO), Melchor Fernández Almagro 3, 28029 Madrid, Spain
e-mail: eballestar@cnio.es

S. Yegnasubramanian and W.B. Isaacs (eds.), *Modern Molecular Biology*, Applied Bioinformatics and Biostatistics in Cancer Research,
DOI 10.1007/978-0-387-69745-1_3,

corresponding target sites. Histone modifications, distribution of nucleasomes and histone variants, transcription factors, and different corepressor and coactivator complexes, among other proteins, define the spatial location of genomic sequences within the nucleus, their transcriptional status, and their timing during replication. Characteristic patterns of epigenetic modifications, chromatin structure, and association of specific transcription factors are associated with established architectural organization and expression patterns that ultimately define the specific features of each cell and tissue type.

If the past decade was marked by the successful completion of the human and mouse genome projects, the present decade has been characterized by the initiation of various projects that will ultimately lead us to understand how genomic information is specifically organized and regulated in each cell type, and how this information can be abnormally used in the context of disease. The development and optimization of a variety of genome-wide strategies has significantly stimulated research interest in the field of epigenomics.

For years, chromatin immunoprecipitation (ChIP) has been one of the most powerful techniques for investigating in vivo interactions between nuclear factors and their genomic target sequences (Orlando 2000; Kuo and Allis 1999). In contrast to DNA footprinting methods, which provide information about the precise target sequence that is bound by a factor or characterized by a particular chromatin structure, in many cases at the nucleotide level, ChIP assays tell us about the nature of the factor, or the histone modification pattern that is bound to a specific DNA segment. In fact, the two techniques are complementary. Although ChIP assays also yield information about the target sequence, it is obtained at a lower resolution than in footprinting experiments. Instead, ChIPs focus on the nature of the bound proteins (histone modification patterns and various factors, including specific coactivators and corepressors, histone modification complexes, chromatin remodeling activities, and other multiprotein complexes) that directly associate with DNA.

In brief, ChIP assays are based on the use of an immunoprecipitating antibody to isolate DNA sequences that are bound by chromatin proteins against which the antibody is raised. After that, immunoprecipitated DNA can be analyzed by standard or quantitative PCR with specific primers by dot blot hybridization to investigate the presence of a candidate DNA sequence. Specific amplification of products demonstrates whether the specific factor or histone modification being interrogated has been bound. Since antibodies can be used to immunoprecipitate both nuclear factors and histone modifications, ChIPs provide dynamic information not only about nuclear factor occupancy at their target binding sites, but also about specific histone modification patterns in selected DNA sequences.

Since ChIP assays theoretically yield all the entire genomic DNA fraction associated with a particular nuclear factor or histone modification pattern, the combination of this technique with hybridization of genomic microarrays (ChIP-chip) allows the generation of maps of distribution of these chromatin features. The availability of a variety of high-resolution microarray platforms has increased the popularity of this strategy, although limitations or complexity associated with the use of bioinformatic tools still make many researchers reluctant to adopt this strategy. More recently, high-throughput sequencing (HTS), also called next generation sequencing, technologies have also become available. The immediate application in the context of chromatin studies is the use of chromatin immunoprecipitation assays to purify DNA fractions and combine them with HTS (ChIP-seq) to obtain high-resolution maps of histone modifications or nuclear factor occupancy (Barski et al. 2007). With this elegant combination of techniques it is now possible to uncover novel binding target sequences for nuclear factors or DNA sequences with specific histone-modification patterns on a genomic scale.

3.2 Experimental Design for a Successful ChIP-Chip or ChIP-Seq Experiment

In order to perform successful ChIP-chip or ChIP-seq experiments several critical considerations need to be taken into account (see Fig. 3.1). Firstly, ChIP conditions need to be optimized and all the requirements for individual ChIP experiments need to be met. The second critical consideration is the amount of DNA necessary for hybridization of microarrays (ChIP-chip) or for analysis with HTS (ChIP-seq). In many this will require non-biased PCR-based random amplification. Finally, for ChIP-chip, the appropriate microarray needs to be carefully selected.

3.2.1 ChIP Assays

To guarantee the success of a ChIP-chip or ChIP-seq experiment it is important to optimize the conditions for the equivalent single-ChIP experiment. Many protocols describing ChIP assays have been published and are now easily available.

There are three critical steps to optimizing and performing an experiment: (a) the covalent fixation of DNA-protein contacts, (b) the generation of chromatin fragments that can be efficiently immunoprecipitated, and (c) the use of appropriate antibodies that are both highly specific and capable of immunoprecipitating a protein that may be partially hindered by chromatin and various types of factors.

ChIP assays generally involve the use of a crosslinking agent that stabilizes protein-DNA contacts; very few protocols make use of native chromatin. The most common crosslinking agent used in ChIP analysis is formaldehyde, a dipolar reagent that produces both protein-nucleic acid and protein-protein crosslinks, involving the amino acids of the imino group, such as Lys, Arg, and His, and DNA (adenines and cytosines). A key property of the crosslinks obtained by using formaldehyde is their reversibility. This can be achieved by treatment at low pH in aqueous solution or incubation at 60–70°C in the presence of sodium dodecyl sulfate (SDS). Due to the small size of the formaldehyde molecule (0.2 nm) only proteins located within this distance of the DNA will become crosslinked. Some of the chromatin-modifying enzymes, such as histone deacetylases or other histone

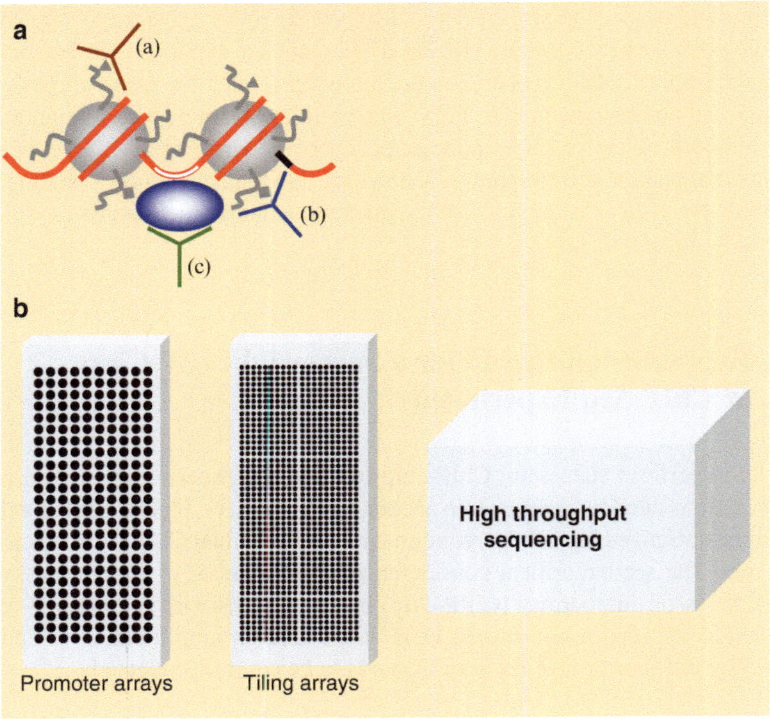

Fig. 3.1 Mapping chromatin at a genome-wide scale by ChIP-chip. (**a**) Chromatin immunoprecipitation provides the basis for isolating DNA sequences with a particular chromatin context. A chromatin fragment is represented. Histone octamers are represented by two *grey spheres*, wrapped by DNA (in *red*). Antibodies to isolate chromatin can be directed against histone modifications (*a*, *b*) at the histone protruding N-terminal tails or a variety of DNA-bound factors (*c*). (**b**) Different types of arrays are currently available: high density promoter arrays (*left*) and tiling arrays (*center*). Tiling arrays exclude repetitive sequences. For full coverage of the entire genome, high throughput sequencing may be necessary

modification enzymes, do not directly bind DNA and their gene-specific regulatory functions operate through recruitment by additional DNA-binding proteins that associate with regulatory sequences. Although these proteins do not exhibit DNA-binding properties, it is possible to investigate their association with particular sequences by using additional protein-protein crosslinkers (Kurdistani and Grunstein 2003). For instance, dimethyl adipimidate has been used to investigate the association with the yeast HDAC Rpd3 (Kurdistani et al. 2002).

Efficient fixation of proteins to DNA is crucial for the ChIP assay. Standard conditions for formaldehyde crosslinking usually consist of a concentration of 1% and an incubation time of around 15 min. However, longer incubation times might be required, depending on the particular chromatin context. At any rate, it is important to avoid long formaldehyde crosslinking treatments as this increases resistance to fragmentation by sonication and decreases the efficiency of the technique.

Moreover, formaldehyde is a moderately strong denaturing agent for proteins. A high concentration or long exposure to this reagent may result in the loss of antigen epitopes. It is advisable to determine empirically the effects of formaldehyde on the protein under study. For instance, after standard fixing conditions for different exposure times, immunolocalization analysis can detect loss of fluorescence signal due to denaturation.

When choosing fixation conditions, it is important to ensure that the increased mechanical resistance of chromatin still allows fragmentation by sonication. This relates to the second critical consideration when performing ChIP assays – the generation of chromatin fragments of appropriate size. This determines the yield of immunoprecipitated material and the degree of resolution of the technique. Chromatin fragmentation is generally achieved by sonication (although micrococcal nuclease digestion provides an alternative method for fragmentation that can be used in native ChIP protocols, i.e., those in which the formaldehyde fixation step is missing) and conditions must be optimized for each sonicator before doing any immunoprecipitation experiment.

In many studies, accurate mapping can be achieved by designing primers that amplify DNA fragments of 200–300 bp.

Specific immunoprecipitation is less efficient with large chromatin fragments than with small fragments. In addition, the size of the fragments determines the resolution of the technique and, therefore, fragments should not greatly exceed the size of the sequence to be analyzed. If the average chromatin fragment is much larger than the sequence to be PCR-amplified or probed, it is not possible to be certain that the protein for which the antibody was used is bound to that particular region or to a neighboring region.

Finally, the quality of the antibody is extremely important in ChIP assays. It is essential to ensure, firstly, that the antibody efficiently recognizes the antigen and, secondly, that most of the immunoprecipitated material represents specific DNA sequences. Ideally, a "no-antibody" control and pre-immune serum control should be included.

We have obtained the best results by using a protocol based on that described by Spencer et al. (2003). The application of such a protocol requires around 1–2 million cells for each antibody. Firstly, formaldehyde is added directly to culture medium to a final concentration of 1%. Normally, 15–30 min is sufficient time to ensure the proper crosslinking (as previously determined) at room temperature. In any case, preliminary experiments to estimate the best combination of crosslinking time and fragmentation should be performed. When planning to store crosslinked cells, glycine should be added to a final concentration of 0.125 M and incubated for 5 min. Once crosslinked and washed with phosphate-buffered saline solution, cells can be directly scraped (in the case of adherent cells) and/or sedimented. In general, crosslinked cell pellets are resuspended in an SDS-containing buffer (typically 1% SDS, 10 mM EDTA, 50 mM Tris-HCl, pH 8.1) for a complete cell lysis that renders denatured chromatin suitable for sonication, thereby giving rise to homogeneous fragmentation. Recently, a bath sonicator has become available for simultaneous, reproducible and effective sonication of multiple samples. To optimize sonication, a time course of conditions should be developed, followed by reversal of crosslinks at 65°C for 4 h. DNA can then be extracted by standard procedures and analyzed in agarose gels to visualize shearing efficiency.

Once sonication conditions have been optimized, soluble chromatin is isolated by centrifugation. At this point, the DNA-containing samples should be approximately quantified. This is particularly advisable when multiple samples are analyzed in order to maintain a constant ratio of chromatin to antibody. Samples are then diluted tenfold and brought to an identical DNA concentration with a dilution buffer (0.01% SDS, 1.1% Triton-X-100, 1.2 mM EDTA, 16.7 mM Tris-HCl, pH 8.1, 167 mM NaCl). Under this low SDS concentration, antibodies can interact and form complexes with their specific antigens. A portion of the diluted cell pellet suspension can be kept to assess the amount of DNA present in different samples for the PCR protocol. This sample is considered to be the input or starting material, and must be heated at 65°C for 4 h in order to reverse crosslinks. To reduce non-specific background, it is advisable to treat the diluted cell lysate with commercially available protein A/G-agarose/salmon sperm DNA beads (50% gel slurry in TE buffer, containing sonicated salmon sperm DNA) for 30 min at 4°C with agitation. Then, agarose beads are pelleted and the supernatant fraction is collected.

The immunoprecipitating antibody is then added to the supernatant fraction and incubated overnight at 4°C with rotation. In general, we have found that 100 μg of sonicated chromatin can be efficiently immunoprecipitated with 5–10 μg of most antibodies. For negative controls, no-antibody immunoprecipitation and pre-immune serum precipitation (when available) must be performed. After overnight incubation, protein A/G-agarose/salmon sperm DNA beads are added and incubated for 1 h at 4°C with rotation to collect the antibody-protein complex. Agarose beads are then pelleted by gentle centrifugation and the supernatant is removed and stored. This fraction, which consists of unbound DNA, is needed to make a direct comparison with the antibody-bound fraction. Subsequently, protein A/G agarose beads are washed briefly (generally 5 min per wash) with a series of buffers of different ionic strength: low salt wash buffer (0.1% SDS, 1% Triton X-100, 2 mM EDTA, 20 mM Tris-HCl, pH 8.1, 150 mM NaCl), high salt wash buffer (0.1% SDS, 1% Triton X-100, 2 mM EDTA, 20 mM Tris-HCl, pH 8.1, 500 mM NaCl), LiCl wash buffer (0.25 M LiCl, 1% NP40, 1% deoxycholate, 1 mM EDTA, 10 mM Tris-HCl, pH 8.1), 1X TE (10 mM Tris-HCl, 1 mM EDTA, pH 8.0). After the last wash, protein-DNA complexes can be eluted from the antibody by adding freshly made elution buffer (1% SDS, 0.1 M $NaHCO_3$) to the pelleted protein A/G agarose-antibody-protein-DNA complex. Generally, two consecutive elution steps are performed and supernatant fractions are pooled. As mentioned above, protein–DNA crosslinks are reversed by heating at 65°C for 4 h. Input samples and the unbound samples are processed in a similar manner. They are then treated with proteinase K and their DNA is extracted by using standard procedures. As indicated above, input, unbound, and bound fractions are commonly analyzed using standard or quantitative PCR with primers designed for a particular genomic region. In general, enrichment of a particular sequence in the bound fraction with respect to the unbound or input fraction indicates the presence of the particular factor or histone modification against which the antibody has been used. When performing ChIP-chip or ChIP-seq experiments, known targets should be individually validated. Typical enrichments (bound versus unbound or input) range from two- to fivefold.

3.2.2 Obtaining Material for Microarray Hybridization or HTS

A single ChIP sample does not usually provide enough DNA for hybridization to a genomic array. Whereas single ChIPs yield DNA amounts in the nanogram range, ChIP-chip experiments require DNA amounts in the microgram range. While the nanogram quantities of DNA obtained after ChIP may sometimes be sufficient to enter HTS workflows, individual ChIP experiments may sometimes yield insufficient amounts for ChIP-seq experiments as well. To obtain the required quantity, multiple ChIP experiments should ideally be performed. It is best to conduct multiple single-standard ChIP assays rather than amplifying the volume in a single experiment. Usually, up to 30 single IP experiments should yield enough material for hybridization for ChIP-chip experiments. Fewer single IP experiments will typically be required for analysis with ChIP-seq. Samples should be treated and processed separately, and only after DNA samples have been resuspended in water should they be pooled before proceeding with fluorescent labeling and hybridization. Obviously, in many cases, only limited amounts of cells or tissues are available, and multiple ChIP experiments require large amounts of antibodies. These two circumstances have highlighted the need to develop different amplification procedures. With these methods, the absence of bias is as important as the capability for generating enough DNA for labeling and hybridization. Several approaches have been taken to overcome this limitation and to obtain the required amounts of DNA.

A commonly used technique for amplifying DNA obtained from ChIP assays is ligation-mediated PCR (LM-PCR). This method is based on the ligation-based addition of a unique DNA linker to the immunoprecipitated chromatin fragments. The addition of an oligonucleotide complementary to this linker allows exponential amplification of any fragment of DNA. Unfortunately, although little bias should be introduced, in practice this method often produces a very strong background when samples are analyzed on genomic arrays (O'Geen et al. 2006).

An alternative method is based on the use of two consecutive amplification steps with a degenerate primer (Kuukasjarvi et al. 1997; Huang et al. 2000). The first step requires the use of thermosequenase, a degenerate oligonucleotide primer (DOP) (5′-CCG ACT CGA GNN NNN NAT GTG G-3′) and low-stringency amplification conditions (3 min at 94°C followed by four cycles of 1 min at 94°C, 1 min at 25°C, 3 min transition at 25–74°C, 2 min extension at 74°C, and a final extension of 10 min). The second step consists of a more standard PCR amplification, standard Taq polymerase, and more stringent conditions (3 min at 94°C, followed by 35 cycles of 1 min at 94°C, 1 min at 56°C, 2 min extension at 72°C, and a final extension of 10 min). Negative controls should be added for each DOP-PCR step in order to discard the existence of non-specific amplification of contaminant DNA.

A third method, know as whole genome amplification (WGA), is based on the isothermal reaction of the enzyme Phi29 and random primers (Blanco et al. 1989). This has become one of the most popular methods currently used in genomic research. At present, there are several commercially available kits that allow it to be carried out simply and efficiently. A recent comparison of amplification methods for ChIP-chip (O'Geen et al. 2006) showed that the signal-to-noise ratio obtained from hybridizations of the WGA products is superior to that from LM-PCR-based methods.

Before labeling and hybridizing the ChIP samples, it is advisable to test a small aliquot of the antibody-treated and no-antibody samples for PCR amplification of positive and negative controls. In general, amplification procedures result in a loss of enrichment of the sequence of interest in the bound fraction. Availability of a known target for the nuclear factor or histone modification of interest helps to validate and confirm that the amplification step has not introduced bias.

3.2.3 Labeling and Hybridizing the DNA for ChIP-Chip

Once the required amount of DNA (1–2 μg) has been obtained for both the antibody ChIP sample and the no-antibody control, it is labeled with the fluorescent Cy5 and Cy3 dyes. There are several commercially available DNA labeling systems for incorporating these dyes into the DNA samples. Once the labeled samples have been obtained, DNAs are cohybridized to the selected microarray. Following hybridization, the arrays are washed, scanned and analyzed like other types of microarray. Many institutions have established core-facility units specialized in microarray hybridization and it is advisable to use their expertise during the analysis.

3.2.4 Choosing the Right Microarray

The possibility of performing ChIP-chip has depended on the availability of suitable genomic microarrays. Several genomic microarray platforms have become available in recent years. It is important to distinguish between those that have a spotted selection of genomic sequences and those with a broad representation of the entire genome. Originally, genomic microarrays were designed for comparative genomic hybridization (CGH) analysis. In this case, large-insert genomic clones, such as bacterial artificial chromosomes (BACs), are used for array spots (Ishkanian et al. 2004). Although this type of microarray can be used in the ChIP-chip technique, the large size of the BAC clones makes it difficult to identify the target sequence of the nuclear factor. Once a positive spot has been identified, additional studies are needed to map the target sequence at a higher resolution within the BAC clone.

More recently, an interesting specialized genomic microarray consisting of a library of CpG island clones was designed by Tim Huang (Yan et al. 2002). This microarray has been used in combination with a method known as differential methylation hybridization. Linker-ligated genomic DNA is digested with a methylation-sensitive restriction enzyme, amplified by PCR, and hybridized to the array. Many CpG islands become methylated in cancer and are thereby protected from methylation-sensitive restriction cleavage and so can be amplified by PCR, producing array-hybridization signals (Huang et al. 1999; Paz et al. 2003). Since CpG islands generally coincide with the promoter of many genes, a CpG island microarray can be useful for investigating the binding sites at the regulatory regions of CpG island-containing genes (Weinmann et al. 2002). We have used Huang's CpG-island microarray to combine with ChIP assays performed with antibodies against methyl-CpG binding domain (MBD) proteins to identify hypermethylated CpG islands in breast cancer cells (Ballestar et al. 2003).

A variety of genomic arrays are currently available. Both promoter-based arrays and whole-genome tiling arrays are commonly used for ChIP-chip experiments. Currently available promoter arrays have a relatively high number of oligonucleotide probes covering the proximal promoter and transcription start site of most human transcripts. This type of array is useful to define the chromatin features of the promoter region of genes at a genome-wide level and to investigate the set of targets of particular events. On the other hand, whole genome tiling arrays contain coding and non-coding regions. Although, this type of array can also be used for ChIP-chip studies they are generally used for high resolution CGH experiments. Despite the great interest in these types of microarrays given their potential use in studying the binding of factors to regulatory regions and of histone modification patterns, repetitive sequences are not represented. For an analysis requiring inclusion of chromatin features in these repetitive sequences, ChIP-seq must be used (Barski et al. 2007).

3.2.5 ChIP-Seq

As introduced above, another option for genome-wide analysis of ChIP enriched DNA is the use of massively parallel HTS (Barski et al. 2007). Here, instead of hybridizing to a microarray, ChIP enriched DNA can be processed into HTS libraries and analyzed using one of the HTS platforms, such as Illumina's Genome Analyzer (see http://www.illumina.com/) or Life Technologies' SOLiD (see http://www.appliedbiosystems.com), each capable of sequencing short reads (25–100 bp) from the ends of millions of DNA molecules from the HTS library in a single run (described in more detail in Chap. 6). The strengths of such a ChIP-seq approach are: (1) ability to interrogate ChIP enrichment sites without any bias to sequence (e.g. does not exclude repeat regions); (2) very low background signals because issues of cross-hybridization, etc. that can affect microarray analyses are not relevant; (3) determination of ChIP enrichment sites with very high resolution because of the high number of sequences generated at each site combined with low surrounding background; and (4) potentially simpler computational and bioinformatics analytical solutions due to the "digital" nature of ChIP-seq (enrichment signals are proportional to "digital" counts of reads sequenced from a given genomic region in ChIP-seq) as opposed to the noisy "analog" nature of ChIP-chip (enrichment signals are proportional to an "analog" microarray hybridization intensity at a given genomic locus). However there are still some disadvantages that have so far limited use of ChIP-seq in favor of ChIP-chip. First, the HTS platforms are still not ubiquitously available, while microarray platforms have become widely accessible to many research centers. Also, although HTS costs are steadily decreasing, ChIP-seq remains considerably more expensive than ChIP-chip when promoter or CpG island arrays are used. For these reasons, both ChIP-seq and ChIP-chip experimental designs will likely remain at the forefront of genome-wide analysis of functional DNA-protein interactions.

3.2.6 Validating ChIP-Chip and ChIP-Seq Results

A key step when using any type of genome-wide discovery analysis is the independent validation of the results. Just as RT-PCR is used to validate the results of genome-wide gene expression analysis, in the case of ChIP-chip or ChIP-seq experiments, individual single-ChIP assays should be performed to confirm the target sequences identified by these technique. It would be ideal to perform ChIPs with two different antibodies raised against the same protein. Specialized validating experiments are advisable. For instance, when we performed ChIP-chip analysis to investigate MBD targets in breast cancer cells (Ballestar et al. 2003), we validated the results by using individual ChIP assays and a specific assay. In this case, since MBDs had been proposed to associate specifically with methylated DNA (Fraga et al. 2003a), we investigated the methylation status of the CpG islands that each of the anti-MBD antibodies had been able to isolate. The specific methylation profile of each of the identified targets was an independent test that served not only to validate the results from the ChIP-chip analysis but also to reveal novel targets of epigenetic inactivation in human breast cancer. For nuclear factors that have a known or inferred binding site, it would be useful to search for that particular binding site in the positive clones resulting from the ChIP-chip experiment. Additionally, electrophoretic mobility-shift experiments can be used to test in vitro the ability to bind the resulting targets (Fraga et al. 2003b).

3.3 ChIP-Chip and ChIP-Seq: When Structural and Functional Information About Chromatin Goes Genome-Wide

In the past 5 years, ChIP-chip and ChIP-seq studies have proliferated revealing the complexity of distribution of histone modification marks and chromatin factors. High resolution maps of histone modifications have successfully been obtained for yeast

(Pokholok et al. 2005), Drosophila (Beisel et al. 2007), mouse and human (Guenther et al. 2007; Barski et al. 2007) cells. In other group of studies the genome-wide distribution of different chromatin factors including co-activators and co-repressors, hormone receptors among other factors has been studied (Zeitlinger et al. 2007; Liu et al. 2008) Different studies reveal the intricate relationship between different epigenetic marks. For instance, recent genome-wide studies on the distribution of Polycomb group genes in normal and cancer cells (Schlesinger et al. 2007; Widschwendter et al. 2007) have shown that genes methylated in cancer cells are specifically packaged with nucleosomes containing histone H3 trimethylated on Lys27. By looking at this chromatin mark in different cell types, it has been shown that trimethyl Lys 27 of H3 is established on unmethylated CpG island genes early in development and then maintained in differentiated cell types by the presence of an EZH2-containing Polycomb complex. In cancer cells, as opposed to normal cells, the presence of this complex brings about the recruitment of DNA methyltransferases, leading to de novo methylation. This example only highlights the complex relationship between epigenetic marks and how genome-wide studies reveal the need for the investment of great research efforts to integrate all the massive information that will be generated.

3.4 Summary

ChIP-chip and ChIP-seq are powerful tools that can be used to discover novel target sequences for transcription factors or to reveal DNA sequences with particular chromatin features. This potential is the result of the elegant combination of ChIP assays with microarray or HTS technology. ChIP assays allow the isolation of a genomic library of sequences that are bound by a specific factor or that contain specific histone modifications. Microarray and HTS technologies make it possible to analyze thousands to millions of sequences in a single experiment. The combination of ChIP-chip and ChIP-seq with other types genome-wide strategies, such as expression microarrays, will surely help to lead to a functional understanding of the way by which the genome is regulated. ChIP-chip and ChIP-seq experiments seem likely to contribute greatly to the mapping of the epigenomic landscape.

References

Ballestar E, Paz MF, Valle L, Wei S, Fraga MF, Espada J, Cigudosa JC, Huang TH, Esteller M (2003) Methyl-CpG binding proteins identify novel sites of epigenetic inactivation in human cancer. EMBO J 22:6335–6345.

Barski A, Cuddapah S, Cui K, Roh TY, Schones DE, Wang Z, Wei G, Chepelev I, Zhao K (2007) High-resolution profiling of histone methylations in the human genome. Cell 129:823–837.

Beisel C, Buness A, Roustan-Espinosa IM, Koch B, Schmitt S, Haas SA, Hild M, Katsuyama T, Paro R (2007) Comparing active and repressed expression states of genes controlled by the Polycomb/Trithorax group proteins. Proc Natl Acad Sci U S A 104:16615–16620.

Blanco L, Bernad A, Lázaro JM, Martín G, Garmendia C, Salas M (1989) Highly efficient DNA synthesis by the phage phi 29 DNA polymerase. Symmetrical mode of DNA replication. J Biol Chem 264:8935–8940.

Fraga MF, Ballestar E, Montoya G, Taysavang P, Wade PA, Esteller M (2003a) The affinity of different MBD proteins for a specific methylated locus depends on their intrinsic binding properties. Nucleic Acids Res 31:1765–1774.

Fraga MF, Ballestar E, Esteller M (2003b) Capillary electrophoresis-based method to quantitate DNA-protein interactions. J Chromatogr B Analyt Technol Biomed Life Sci 789:431–435.

Guenther MG, Levine SS, Boyer LA, Jaenisch R, Young RA (2007) A chromatin landmark and transcription initiation at most promoters in human cells. Cell 130:77–88.

Huang TH, Perry MR, Laux DE (1999) Methylation profiling of CpG islands in human breast cancer cells. Hum Mol Genet 8:459–470.

Huang Q, Schantz SP, Rao PH, Mo J, McCormick SA, Chaganti RS (2000) Improving degenerate oligonucleotide primed PCR-comparative genomic hybridization for analysis of DNA copy number changes in tumors. Genes Chromosomes Cancer 28:395–403.

Ishkanian AS, Malloff CA, Watson SK, DeLeeuw RJ, Chi B, Coe BP, Snijders A, Albertson DG, Pinkel D, Marra MA, Ling V, MacAulay C, Lam WL (2004) A tiling resolution DNA microarray with complete coverage of the human genome. Nat Genet 36:299–303.

Kuo MH, Allis CD (1999) In vivo cross-linking and immunoprecipitation for studying dynamic Protein:DNA associations in a chromatin environment. Methods 19:425–433.

Kurdistani SK, Grunstein M (2003) In vivo protein-protein and protein-DNA crosslinking for genomewide binding microarray. Methods 31:90.

Kurdistani SK, Robyr D, Tavazoie S, Grunstein M (2002) Genome-wide binding map of the histone deacetylase Rpd3 in yeast. Nat Genet 31:248.

Kuukasjarvi T, Tanner M, Pennanen S, Karhu R, Visakorpi T, Isola J (1997) Optimizing DOP-PCR for universal amplification of small DNA samples in comparative genomic hybridization. Genes Chromosomes Cancer 18:94–101.

Liu Y, Gao H, Marstrand TT, Ström A, Valen E, Sandelin A, Gustafsson JA, Dahlman-Wright K (2008) The genome landscape of ERalpha- and ERbeta-binding DNA regions. Proc Natl Acad Sci U S A 105:2604–2609.

O'Geen H, Nicolet CM, Blahnik K, Green R, Farnham PJ (2006) Comparison of sample preparation methods for ChIP-chip assays. Biotechniques 41:577–580.

Orlando V (2000) Mapping chromosomal proteins in vivo by formaldehyde-crosslinked-chromatin immunoprecipitation. Trends Biochem Sci 25:99–104, Review.

Paz MF, Wei S, Cigudosa JC, Rodriguez-Perales S, Peinado MA, Huang TH, Esteller M (2003) Genetic unmasking of epigenetically silenced tumor suppressor genes in colon cancer cells deficient in DNA methyltransferases. Hum Mol Genet 12:2209–2219.

Pokholok DK, Harbison CT, Levine S, Cole M, Hannett NM, Lee TI, Bell GW, Walker K, Rolfe PA, Herbolsheimer E, Zeitlinger J, Lewitter F, Gifford DK, Young RA (2005) Genome-wide map of nucleosome acetylation and methylation in yeast. Cell 26:517–527.

Schlesinger Y, Straussman R, Keshet I, Farkash S, Hecht M, Zimmerman J, Eden E, Yakhini Z, Ben-Shushan E, Reubinoff BE, Bergman Y, Simon I, Cedar H (2007) Polycomb-mediated methylation on Lys27 of histone H3 pre-marks genes for de novo methylation in cancer. Nat Genet 39:232–236.

Spencer VA, Sun JM, Li L, Davie JR (2003) Chromatin immunoprecipitation: a tool for studying histone acetylation and transcription factor binding. Methods 31:67.

Weinmann AS, Yan PS, Oberley MJ, Huang TH, Farnham PJ (2002) Isolating human transcription factor targets by coupling chromatin immunoprecipitation and CpG island microarray analysis. Genes Dev 16:235–244.

Widschwendter M, Fiegl H, Egle D, Mueller-Holzner E, Spizzo G, Marth C, Weisenberger DJ, Campan M, Young J, Jacobs I, Laird PW (2007) Epigenetic stem cell signature in cancer. Nat Genet 39:157–158.

Yan PS, Efferth T, Chen HL, Lin J, Rodel F, Fuzesi L, Huang TH (2002) Use of CpG island microarrays to identify colorectal tumors with a high degree of concurrent methylation. Methods 27:162–169.

Zeitlinger J, Zinzen RP, Stark A, Kellis M, Zhang H, Young RA, Levine M (2007) Whole-genome ChIP-chip analysis of Dorsal, Twist, and Snail suggests integration of diverse patterning processes in the Drosophila embryo. Genes Dev 21:385–390.

Chapter 4
Genome-Wide DNA Methylation Analysis in Cancer Research

Srinivasan Yegnasubramanian and William G. Nelson

Abstract DNA methylation is a central epigenetic process involved in establishing normal cellular gene expression patterns and genome homeostasis. Aberrations in DNA methylation, leading to abnormal gene expression patterns, have now been linked to many human diseases, and are a nearly universal feature of human cancers. Because these DNA methylation changes can be stably transmitted during clonal outgrowth of cancer cells, they can carry the same importance as mutations in the initiation and progression of human cancers. Such somatic DNA methylation changes often occur earlier and more frequently than genome mutations during carcinogenesis, and have therefore provided a wealth of targets for translational opportunities in cancer biomarkers for diagnosis and risk stratification. Additionally, since these DNA methylation changes are epigenetic processes that are enzymatically mediated and do not alter the underlying DNA sequence, they can potentially be reversed by pharmacological inhibition of the epigenetic machinery, providing opportunities for cancer therapy. Therefore, understanding the genome-wide patterns of DNA methylation in normal and cancer cells has become of primary interest in cancer research. In this chapter, we will first provide an overview of DNA methylation as an epigenetic process in normal physiology and in carcinogenesis. Then we will describe some of the current and upcoming technologies used in analyzing DNA methylation patterns at a genome-wide level, and consider the strengths and limitations of each of these approaches.

4.1 Introduction, Background and Significance

Virtually all somatic cells within any individual contain identical primary genomic DNA sequence information. Yet cells of different lineages, organs, and even different microenvironments within the same organs have vastly differing phenotypes

S. Yegnasubramanian (✉)
Sidney Kimmel Comprehensive Cancer Center, Johns Hopkins University School of Medicine, Baltimore, MD, USA
e-mail: syegnasu@jhmi.edu

S. Yegnasubramanian and W.B. Isaacs (eds.), *Modern Molecular Biology*, Applied Bioinformatics and Biostatistics in Cancer Research,
DOI 10.1007/978-0-387-69745-1_4,

and gene expression profiles. Epigenetics is the study of heritable processes by which cells establish unique gene expression patterns without changing their gene sequence. These epigenetic processes constitute a level of coding beyond the primary sequence and are likely responsible for establishing the spectrum of gene expression changes observed during development and differentiation. Aberrations in these processes appear to be among the earliest and most frequent somatic changes in human cancers, contributing to the initiation of malignant transformation and progression to advanced disease.

4.1.1 DNA Methylation in Physiology and Cancer Pathophysiology

Among the most widely studied epigenetic processes is DNA methylation. In vertebrate genomes, DNA methylation occurs predominantly at the 5-position of cytosine (C) to form 5-methyl-cytosine (5meC) in self-complementary CpG dinucleotides by the activity of DNA methyltransferase (DNMT) enzymes (Jones and Liang 2009). The mammalian DNMTs, which include DNMT1, DNMT3a, and DNMT3b, are central to the establishment and maintenance of DNA methylation during physiological processes as well as during carcinogenesis. These enzymes catalyze the transfer of a methyl group from *S*-adenosyl-methionine to the 5-position of cytosine bases in CpG dinucleotides, although there is evidence to suggest that DNMT3a may promiscuously modify C in non-CpG contexts in embryonic stem cells and various stages of development (Ramsahoye et al. 2000; Gowher and Jeltsch 2001). The methylation of CpG dinucleotides is known to be central to several physiological processes including development, imprinting (Tilghman 1999; Feinberg et al. 2002; Onyango et al. 2002), X-chromosome inactivation (Norris et al. 1991), suppression of repetitive DNA elements (Chapman et al. 1984; Tolberg et al. 1987; Shinar et al. 1989; Challita et al. 1995), and transcription (Razin and Riggs 1980; Siegfried and Cedar 1997; Siegfried et al. 1999).

The role of DNA methylation in transcriptional regulation is perhaps the best studied of these processes. The self-complementary CpG dinucleotide is usually methylated in the normal somatic cell genome, and is highly under-represented compared to all other dinucleotides (Bird 1986). This under-representation presumably occurs because spontaneous hydrolytic deamination of 5meC to thymine in germ cell genomes has led to depletion of CpG dinucleotides during evolution (Bird 1986). Despite this overall under-representation of CpG dinucleotides, dense clusters of CpG dinucleotides, termed CpG islands (CGIs), which are usually unmethylated in normal somatic cell genomes, are found at the transcriptional regulatory regions of >60% of genes (Cross et al. 1994). In the unmethylated state, these CGIs can be housed in active chromatin conformations that are permissive of gene expression.

DNA methylation at these CGIs is associated with condensation of local chromatin by chromatin remodeling complexes in a manner that resembles the facultative heterochromatin seen in the inactive X chromosome in female somatic cells (Bird 1986). This condensed local chromatin structure is resistant to loading of RNA polymerase II and therefore leads to its transcriptional inactivation (Bird 1986). Such DNA methylation induced gene silencing events have long been supposed to mediate tissue- and developmental/differentiation stage-specific gene expression patterns. However, only recently, with the use of unbiased genome-wide methylation detection technologies, have such tissue differentially methylated and expressed genes been identified systematically in mammalian genomes (Song et al. 2005; Lister et al. 2009).

The formation of these condensed chromatin conformations appears to be dependent on the action of specific methyl-DNA binding proteins, such as the methyl-binding domain (MBD) family of proteins (MBD1, MBD2, MeCP2), certain zinc-finger/BTB domain proteins (Kaiso, ZBTB4, and ZBTB8), and a recently identified SRA domain protein (UHRF1) that can preferentially recognize hemi-methylated DNA (Bostick et al. 2007; Sharif et al. 2007). Although the comprehensive mechanisms have not fully been characterized, these proteins can specifically recognize and bind methylated DNA and recruit other chromatin altering proteins and enzymes to facilitate changes in chromatin structure or function. These proteins thus mediate the signal transduction and interpretation of the DNA methylation code.

Much work has focused on the derangement of these physiological DNA methylation processes in the initiation and progression of human malignancies. Early studies examining aberrations in DNA methylation in human cancers showed that cancer genomes have reduced genomic 5meC content compared to normal genomes and also become undermethylated at CpG dinucleotides within the genes (Feinberg and Vogelstein 1983a, b; Gama-Sosa et al. 1983; Goelz et al. 1985; Bedford and van Helden 1987; Feinberg et al. 1988). While the consequences of these changes remain largely unknown, subsequent work has suggested that this hypomethylation may result in genomic instability due to increased rearrangements (Feinberg and Tycko 2004; Cadieux et al. 2006; Rodriguez et al. 2006; Suzuki et al. 2006).

Cancer genomes also appear to harbor abnormal DNA hypermethylation at CGI sequences resulting in an inappropriate silencing of the associated gene (Lee et al. 1994; Esteller et al. 2001). Like gene deletions and mutations, DNA hypermethylation and the resulting epigenetic transcriptional repression has been postulated to be an important means by which cancer cells acquire and maintain their malignant phenotype (Jones and Laird 1999). These findings underline a fundamental enigma in the generation of abnormal DNA methylation patterns in cancer cells, in which there may be a decrease in overall genomic 5meC content with a paradoxical increase in CpG methylation at certain CGIs (Ehrlich 2002). Interestingly, it has been found that DNA hypermethylation at the promoters of genes can cooperate with genetic modes of silencing such as mutations and deletions to satisfy each hit of the Knudson two-hit hypothesis for inactivation of tumor-suppressor genes (Jones and Laird 1999). Conversely, DNA hypomethylation can lead to gene activation and can cooperate with amplifications and mutations for oncogene activation. Therefore, it is becoming increasingly apparent that integrative analysis of DNA methylation and other epigenetic processes along with assessment of genetic processes will be crucial to further our understanding of carcinogenesis and develop rational targeted biomarkers and therapies.

4.1.2 Clinical Translational Potential of Cancer-Associated Somatic DNA Methylation Alterations

Somatic epigenetic alterations, particularly DNA methylation changes, offer a great source of potential molecular biomarkers for cancer diagnosis and risk stratification for several reasons (Laird 2003). First, somatic CGI hypermethylation changes have been nearly universally identified in all human cancers. Second, these somatic CGI hypermethylation changes often appear to be more prevalently associated with cancers than other somatic genetic changes such as mutations, deletions, and translocations. Finally, a number of sensitive and specific strategies are being developed to detect CGI methylation from scant genomic DNA sources such as bodily fluids and biopsy specimens (Sidransky 2002; Laird 2003; Bastian et al. 2005; Li et al. 2009).

Understanding DNA methylation alterations can also provide approaches for rational/targeted cancer therapeutics. Unlike mutations, deletions, translocations, and amplifications, somatic changes in DNA methylation and other epigenetic processes are potentially reversible, making epigenetic genome defects one of the most attractive rational therapeutic targets for treatment of cancer. In support of this, small molecule inhibitors of DNA methyltransferases (DNMTs) and histone deacetylases (HDACs) have already secured US Food and Drug Administration (FDA) approval, as single agents, for myelodysplasia and cutaneous T-cell lymphoma, respectively, and several other epigenetic drugs are under active preclinical and clinical development (Kaminskas et al. 2005; Kantarjian et al. 2006; Mann et al. 2007).

4.1.3 Overview of Approaches for Detection of DNA Methylation

Given the importance of DNA methylation in health and disease, there is great interest in characterizing DNA methylation patterns during carcinogenesis and disease progression. In order to accomplish such analysis, researchers have had to overcome several challenges unique to DNA methylation analysis that do not encumber analysis of genetic mutations (Laird 2010). Because the methylation of cytosines does not alter base pairing and DNA polymerases do not differentiate between C and 5meC during polymerization, DNA methylation patterns cannot be detected by routine molecular biology approaches such as hybridization, PCR or cloning as can be done of genetic mutations that alter the underlying genetic sequence. Instead, researchers most commonly rely on indirect approaches for detection of 5meC. These approaches can be categorized into three main categories: (a) sodium bisulfite conversion of DNA, (b) digestion with methylation sensitive or specific restriction enzymes, or (c) affinity enrichment of methylated DNA. Following processing with these approaches, methylation alterations can be detected by other routine molecular biology approaches, such as Southern blot, gel fractionation, PCR, or sequencing. More recently, these approaches have been coupled with high-throughput platforms such as microarrays or next generation sequencing (NGS) to allow massively parallel genome-wide analysis of DNA methylation patterns in health and disease. In the future, major improvements in single molecule detection strategies may allow direct assessment of DNA methylation across the genome at single base resolution while simultaneously addressing genetic alterations.

4.2 Sodium Bisulfite Conversion Based Methods for DNA Methylation Analysis

One of the most widely used approaches for delineation of DNA methylation patterns features the use of sodium bisulfite to deaminate cytosine to uracil while leaving 5meC intact (Wang et al. 1980). This creates DNA sequence differences at C versus 5meC that can be taken advantage of by routine molecular biology approaches such as hybridization, PCR, cloning, and sequencing (Fig. 4.1). Indeed,

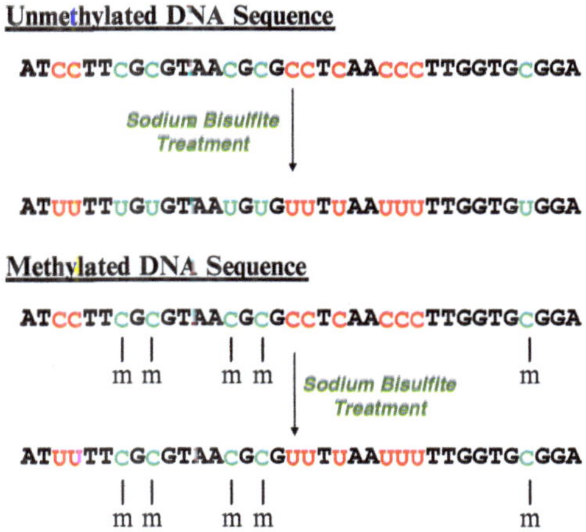

Fig. 4.1 Schematic of sodium bisulfite conversion. Treatment of genomic DNA with sodium bisulfite and subsequent desulfonation under appropriate conditions can allow the specific conversion of unmethylated cytosines to uracil, while methylated cytosines will remain intact. The subsequent sequence difference can be detected by multiple methods including sequencing, PCR, microarray hybridization, etc. Lowercase m denotes methylation of cytosine. Cytosines in CpG dinucleotides are marked in blue

the gold standard approach for locus-specific DNA methylation analysis, now extendable to genome-wide analysis, is bisulfite genomic sequencing (Frommer et al. 1992). In this technique, PCR primers are complementary to the bisulfite converted alleles, but do not overlap with potentially methylated cytosines (those in CpG dinucleotides). PCR amplification, cloning of PCR products into plasmids, and subsequent sequencing will reveal the prevalence of each pattern of CpG methylation in the original sample at single base resolution. More recently, pyrosequencing (Dupont et al. 2004) and base-specific cleavage coupled with MALDI mass spectrometry (Ehrich et al. 2005) has been used to read out bisulfite sequencing of PCR products instead of conventional Sanger sequencing of individual clones. These approaches can be used to quantitatively estimate the fraction of alleles methylated at any given CpG within the PCR amplicon. Another locus-specific bisulfite based strategy, called methylation specific PCR (MSP) (Herman et al. 1996), uses PCR primers targeting the bisulfite induced sequence changes to specifically amplify either methylated or unmethylated alleles, and can be used to detect the presence of a single pattern of methylation in each reaction. Quantitative variations of this technique, such as MethyLight (Eads et al. 2000), HeavyMethyl (Cottrell et al. 2004), RT-MSP (Yegnasubramanian et al. 2004), and MethylQuant (Thomassin et al. 2004), employ methylation specific oligonucleotide primers in conjunction with Taqman probes or SYBR Green based real-time PCR amplification to quantitate loci with a specific pattern of methylation. Sodium bisulfite based methods have been coupled with microarrays and NGS to allow genome-wide analysis of DNA methylation as described below.

4.2.1 Sodium Bisulfite Conversion Coupled with Microarrays for Genome-Wide DNA Methylation Analysis

Highly parallel analysis of the methylation pattern of thousands of CpG dinucleotides across the genome has been made possible by coupling sodium bisulfite conversion of genomic DNA with microarray analysis. Among the first of these strategies essentially featured bisulfite conversion followed by amplification of multiple individual genomic regions and hybridization of these pooled amplicons to microarrays containing probes specific to either the methylated or unmethylated segments of these amplicons (Gitan et al. 2002). These approaches represented modest improvements in target throughput compared to locus-specific bisulfite based approaches described above. As a major extension of these array-based methods, Illumina has commercialized the GoldenGate methylation assay, a multiplex assay that allows simultaneous registration of ~1,500 CpG sites via extension-dependent ligation of multiplexed barcoded primers that are specific to either the methylated or unmethylated bisulfite converted template and subsequent hybridization to the Illumina Bead Array platform (Bibikova and Fan 2009). Since the assay is suited to carry out analysis of 96 samples in parallel, and can assay ~1,500 CpG sites from ~800 genes across the genome simultaneously, and supports the possibility for custom content to interrogate other CpGs of interest, this has become a popular method for highly parallel bisulfite conversion based CpG methylation analysis. More recently, Illumina has extended this approach to the Infinium assay platform, composed of whole-genome amplification of bisulfite converted templates, fragmentation of the amplified library, and hybridization to bead arrays with each methylation-specific probe specific to sequences surrounding a single CpG site (Bibikova et al. 2009). A single-nucleotide extension allows incorporation of a fluorescently labeled base opposite the G in the interrogated CpG dinucleotide only if the DNA template correctly bound to the methylated (C containing) or unmethylated (T containing) probe. This adaptation has allowed tremendous increase in target throughput, with interrogation of >27,000 CpGs from ~14,500 protein-coding and 110 microRNA gene promoters in the current version of the platform. Nonetheless, thus far, all of the microarray based bisulfite conversion methods interrogate a select subset of CpG dinucleotides in the genome, and are therefore not completely unbiased or genome-wide in scope.

4.2.2 Sodium Bisulfite Conversion Coupled with Conventional or Next-Generation Sequencing for Genome-Wide DNA Methylation Analysis

The earliest applications of sequencing for analysis of genome-scale bisulfite conversion based DNA methylation analysis were simply brute force application of individual bisulfite genomic sequencing assays across thousands of amplicons across multiple human chromosomes (Eckhardt et al. 2006). While there was much biological insight gained from these studies, the cost and labor involved in carrying them out was prohibitive for wide adoption.

The advent of next generation sequencing technologies allowed the revisitation of genome-wide bisulfite genomic sequencing approaches. Initial forays in the application of NGS to bisulfite conversion based DNA methylatoin analysis involved ultra-deep sequencing of pooled bisulfite sequencing PCR products, generating >1,000× coverage of each product for highly sensitive DNA methylation analysis for a modest number of targets (Taylor et al. 2007).

The next increase in target throughput scale facilitated by NGS was dependent upon enriching for known portions of the genome for massively parallel bisulfite genomic sequencing rather than attempting shotgun bisulfite sequencing of the entire genome. The first of these targeted massively parallel bisulfite sequencing approaches is called reduced representation bisulfite sequencing (RRBS) (Meissner et al. 2005, 2008). In this method, the genome is fragmented with a methylation insensitive restriction enzyme (thus far, BglII or MspI have been used) that contains a CpG in its target sequence. The resulting fragments are then size selected such that a reproducible portion of the genome, that is highly enriched for CpG rich regions (approximately 1% of all CpGs can be interrogated), is obtained. These fragments are then ligated to next generation sequencing adaptors and subjected to massively parallel short-read sequencing starting at each restriction enzyme site. This approach was shown to be: (a) extendable to tissue specimens, including paraffin embedded formalin fixed tissues (Gu et al. 2010), (b) could be used to assess DNA methylation patterns from as little as 30 ng input material (Gu et al. 2010), and (c) has revealed interesting distinctions between the methylation patterns in embryonic stem cells and their differentiated counterparts (Meissner et al. 2008). Other approaches for bisulfite sequencing of a subset of the genome have used sequence hybridization techniques to capture bisulfite converted DNA fragments followed by massively parallel bisulfite sequencing to register the methylation patterns in those captured fragments (Hodges et al. 2009). Strategies for capture can include hybridization to a library of RNA oligos in solution, hybridization to microrrays, or hybridization dependent ligation of locus-specific padlock probes followed by universal primer amplification (Laird 2010). The disadvantage of such approaches is that the capture probes would have to account for all possible combinations of methylation at cytosines and the sequence differences may cause fluctuations in the degree of capture. An alternative approach would be to capture prior to bisulfite conversion. However, since the efficiency of capture is low, this strategy would often require prohibitively high amounts in the order of dozens of micrograms of input DNA for robust capture followed by bisulfite conversion and sequencing. Nonetheless, this is a promising approach that will surely be used in the future.

More recently, due to significant gains in the sequencing output of next generation sequencing platforms, multiple groups have been able to carry out whole genome shotgun bisulfite sequencing. The major drawback to such a strategy is that the cost remains somewhat prohibitive since significant coverage (50–100×) of each base in the genome is needed to facilitate high-confidence alignment of the bisulfite converted DNA to the reference genome(Lister et al. 2009). Even with this extremely high coverage, some 10% of CpGs in the genome cannot be interrogated because their surrounding sequence would not be unique after bisulfite conversion (Laird 2010).

4.2.3 Analytical Considerations for Bisulfite Sequencing Based Approaches

The major analytical issues encountered in bisulfite-conversion based approaches is that of reduced sequence complexity. The process of bisulfite conversion essentially turns the four-base genome into a three-base genome devoid of C except at the positions of methylated cytosines. The resulting converted sequence poses several challenges for both microarray and sequencing based approaches. For microarrays, the major bisulfite conversion related issue is that of increased cross hybridization of target DNA to imperfectly paired probes on the microarray due to the reduced sequence complexity. Furthermore, since the converted template DNA molecules often have poor G+C content, the affinity of these targets to their capture probes may be low. Additionally, there may be accentuated probe effects between probes specific for methylated and unmethylated targets, with the methylated probes inherently more "sticky" than the unmethylated ones due to the increased G+C content in these probes. Therefore analytical approaches must account for the resulting fluctuations in signal to noise ratio. Often, this may involve empirically discarding poorly performing probes from analysis as has been done for the Illumina methylation platforms (The Cancer Genome Atlas Project 2008). For next generation sequencing, the primary concern is that of alignment. The decreased sequence complexity makes alignment challenging and computationally intensive due to the need to account for all the possible combinations of methylated and unmethylated and converted and unconverted cytosines in any given read. Additionally, there will be some CpGs for which it is even theoretically impossible to measure the methylation level because the surrounding sequence context cannot be uniquely mapped due to the reduction in complexity following bisulfite treatment. As read lengths increase and mate-paired library approaches are implemented, these difficulties should diminish.

4.3 Methylation-Sensitive and -Specific Restriction Endonuclease (MSRE) Based Methods for DNA Methylation Analysis

Another approach to detection of DNA methylation patterns takes advantage of the property of certain restriction endonucleases to fail to cut target sequences methylated at CpG dinucleotides (e.g. HpaII or SmaI) or to specifically cut target sequences harboring methylated CpG dinucleotides (e.g. McrBc or GlaI). In principle, such approaches can be used to enrich for methylated DNA or for unmethylated DNA depending on the processing strategy used (Fig. 4.2). The methylation sensitive restriction enzymes are used more frequently than the methylation specific ones because they have been available commercially for longer, and have isoschizomers that do are not sensitive to methylation (e.g., MspI for HpaII, and XmaI for SmaI) allowing for robust control experiments to ensure that the target site was not mutated in the sample. The most common applications using the methylation sensitive restriction enzymes feature digestion of unmethylated DNA while leaving methylated DNA intact for detection by Southern blot analysis (Bird and Southern 1978; Singer et al. 1979; Pollack et al. 1980), PCR or real-time PCR using primers flanking the target site (Singer-Sam et al. 1990a, b; Bastian et al. 2005). The Southern blot strategy is not easily amenable to high throughput analysis, and requires copious amounts of high molecular weight DNA. Digestion followed by PCR is sensitive, but is limited to interrogating methylation only at the enzyme recognition sites and is plagued by a propensity for false-positives resulting from incomplete digestion on unmethylated loci. The first efforts for genome-scale comparative methylation analysis featured digestion with methylation sensitive restriction enzymes and resolution of the resulting fragments by two

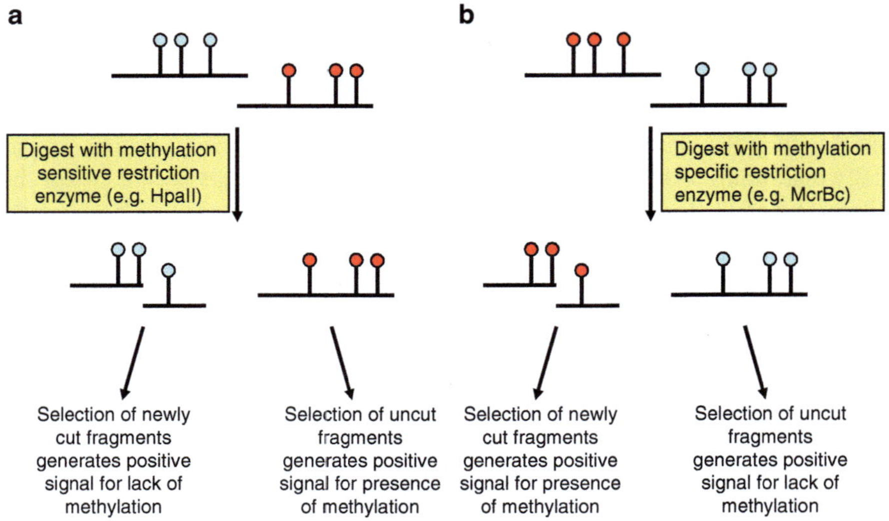

Fig. 4.2 Schematic of potential strategies for DNA methylation analysis using methylation-sensitive of methylation-specific restriction enzymes. (**a**) Methylation sensitive restriction enzymes (e.g. HpaII) cut only when their target sequence is unmethylated. Selection of cut fragments by linker ligation to newly cut ends and amplification will allow positive selection of unmethylated sequences. Selection by amplification with primers flanking the target site will allow generation of positive signals for methylated sequences. (**b**) Use of methylation specific restriction enzymes (e.g. McrBc) leads to digestion of DNA only when the target sequence is methylated. In this case, selection of cut fragments will lead to positive signals for methylated DNA, and selection of uncut fragments will lead to positive signals for unmethylated DNA. Red "lollipops" denote methylated CpGs and blue "lollipops" denote unmethylated CpGs

dimensional gel electrophoresis. Comparison of patterns between two samples can allow identification of fragments that were differentially methylated between the two. The identification of the sequence underlying the differential patterns elucidated by gel electrophoresis can be accomplished by sequencing of the fragments. This method, called Restriction Landmark Genome Scanning (RLGS) (Hayashizaki et al. 1993), has yielded valuable information on genome wide DNA methylation patterns including elucidation of tissue-differentially methylated regions (Song et al. 2005), imprinted regions (Plass et al. 1996), and cancer specific differentially methylated regions (Costello et al. 2000). Other approaches involving MSRE digestion and resolution by gel electrophoresis include amplification of intermethylated sequences (AIMS) (Frigola et al. 2002) and methylation-sensitive arbitrarily primed PCR (MS-AP-PCR) (Liang et al. 2002). In these methods, the digested material is subjected to an arbitrarily primed PCR, and differential product sizes are discriminated by gel electrophoresis. As microarray and NGS approaches have developed, these gel-based approaches have become less commonly used.

4.3.1 MSRE Fractionation Coupled with Microarrays or NGS for Genome-Wide DNA Methylation Analysis

Differential methylation hybridization (DMH) was one of the first robust MSRE-microarray based approaches for genome-wide methylation analysis (Yan et al. 2009). In this approach, the DNA is first cut with the MseI restriction enzyme, which cuts at AATT sequences, and then adapted to amplification linkers. MseI is a highly frequent cutter in most of the genome, but cuts relatively infrequently at CGIs, allowing some degree of enrichment for CGI sequences. Next, the DNA is split into two pools and each is subjected to digestion with a methylation sensitive restriction enzyme or to a mock digestion in which the enzyme is omitted. Each pool is then amplified with primers specific to the universal adaptors and labeled with different fluorescent dyes. The resulting products are then combined together and hybridized to two-channel microarrays. The ratio of fluorescence from the MSRE digested pool to the fluorescence from the undigested pool will provide the degree of methylation at each locus interrogated on the array.

Another assay, called HpaII tiny fragment enrichment by ligation mediated PCR (HELP), involves digestion of genomic DNA with HpaII (methylation sensitive) or MspI restriction enzymes in separate pools, ligating adaptors to the newly generated ends, amplifying with adapter specific primers and labeling, and pooling and hybridizing to microarrays (Khulan et al. 2006). The HELP assay has also been adapted for analysis with NGS (Oda et al. 2009). The HELP method generates a positive signal for unmethylated DNA since the adaptor ligation and amplification is possible only at unmethylated HpaII target sites. Other variations of this overall strategy such as Methyl-Seq (Brunner et al. 2009) and methylation-sensitive cut counting (MSCC) (Ball et al. 2009) have also been used with NGS.

Other variations of restriction enzyme based methods include those that employ the methylation-specific restriction enzyme McrBc. MethylScope was one of the first such methods to do this coupled with microarrays (Ordway et al. 2006). The CHARM assay built upon this by introducing improvements in microarray design and microarray data analysis (Irizarry et al. 2008). In another approach, called MMASS, samples are split into two fractions, and the first is digested with a pool of methylation sensitive restriction enzymes and the second is digested with a methylation-specific restriction enzyme (McrBc) and the two pools are compared (Ibrahim et al. 2006). In principle, these approaches should also be adaptable to analysis with NGS instead of microarrays.

4.4 Affinity Enrichment Based Methods for Genome-Wide DNA Methylation Analysis

4.4.1 Affinity Reagents for Recognition of Methylated DNA

Another approach for detection of DNA methylation features the use of proteins with selective affinity for methylated DNA compared to unmethylated DNA. One of the first examples, introduced in 1994 by Cross et al. used column-immobilized recombinant methylated-CpG binding domain (MBD) proteins to enrich for methylated DNA fragments (Cross et al. 1994). Since then, more recent approaches have exploited enrichment of methylated DNA using antibodies specific for methylcytosine residues (5meC-Ab) (Weber et al. 2007), the MBD of MBD1(Jorgensen et al. 2006), MBD2 (Gebhard et al. 2006b; Yegnasubramanian et al. 2006), or MeCP2 (Zhang et al. 2006), or full length MBD containing proteins(Rauch et al. 2008). Of these different 5meC affinity reagents, the MBD of MBD2 (MBD2-MBD) appears to have the highest affinity for a wide range of methylated DNA sequences and doesn't seem to have preference for any specific consensus sequences or sequence motifs outside of the 5meCpG dinucleotide sequence. MECP2, on the other hand, may more selectively bind to 5meCpG dinucleotides adjacent to A/T rich sequences (Klose et al. 2005). We and others have also shown that the MBD2-MBD can be used to capture a large variety of methylated CGI sequences (Yegnasubramanian et al. 2006). The 5meC-Ab is less ideal because, unlike the MBD2-MBD which can bind double-stranded methylated DNA, it can only bind single stranded DNA. This is particularly problematic since the high G/C content of CGIs may make these sequences resistant to denaturing and prone to forming secondary structures even after denaturing. Additionally, using just the small ~10 kD MBD portion of the MBD2 protein, as opposed to the full length protein, could improve sensitivity and specificity for methylated DNA. To overcome some of the limitations, groups have concatamerized MBD domains from other MBD family members for improved affinity (Jorgensen et al. 2006), and have used full-length MBD2 proteins along with MBD3L proteins (e.g. methylated CpG island recovery assay or MIRA), resulting in increased affinity of the full-length protein (Rauch and Pfeifer 2005). For gene-specific DNA methylation analysis, these enrichment strategies can be coupled with PCR or real-time PCR to assess whether a given genomic locus is methylated (Rauch and Pfeifer 2005; Gebhard et al. 2006a; Yegnasubramanian et al. 2006). However, non-specific binding of unmethylated DNA to the capture substrate or to the protein itself can result in false positive results (Yegnasubramanian et al. 2006). To prevent this, a combination of methylated DNA precipitation with methylation sensitive restriction enzyme (COMPARE-MS) approach has been used to improve the sensitivity and specificity of gene-specific affinity-enrichment based methylation detection (Yegnasubramanian et al. 2006).

4.4.2 Affinity-Enrichment of Methylated DNA Coupled with Microarrays or NGS for Genome-Wide DNA Methylation Analysis

The affinity-based DNA methylation enrichment approaches have now been coupled with microarray and NGS analysis for characterization of DNA methylation patterns genome-wide (Fig. 4.3). Essentially, genomic DNA is fragmented to small

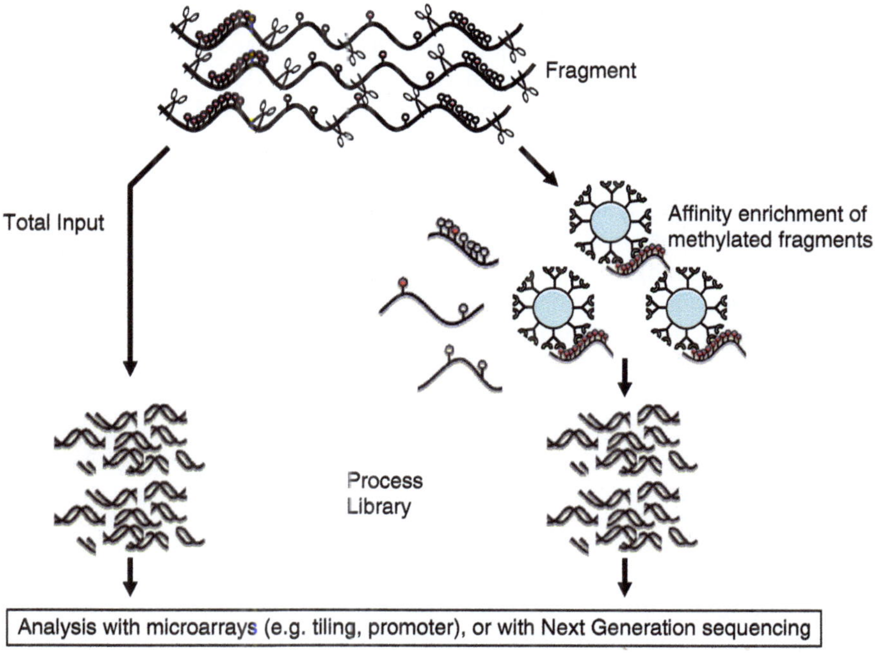

Fig. 4.3 Schematic of general approach for genome-wide methylation analysis using affinity reagents for enrichment of methylated DNA. Genomic DNA is fragmented and divided into two pools. The first pool is untreated and represents the total input. The second pool is subjected to affinity enrichment using proteins that specifically bind methylated DNA (e.g., 5meC-Ab, or MBD proteins) that are immobilized on beads or columns. Each pool is then subjected to appropriate library generation steps for analysis with either microarrays or NGS. *Blue circles* denote particles on which affinity reagents, such as the 5meC-Ab are immobilized. *Red circles* denote methylated CpGs. *Clear circles* denote unmethylated CpGs

sizes (typically 100–1,000 bp) and then split into two fractions. One fraction is subjected to enrichment for methylated DNA fragments by binding to affinity reagents immobilized on a column or on beads. The other fraction is left as an unprocessed total input. Equal amounts of each of these fractions are then processed for microarray hybridization or analysis by NGS. Comparison of the enriched methylated fraction with the total input allows the ability to discriminate methylated regions. There are several such methods that have been reported. Most of these are highly similar to each other, with the major difference being the enrichment reagent. Use of the 5meC-Ab coupled with microarrays or NGS has been referred to as MeDIP or MeDIP-seq (Weber et al. 2007; Down et al. 2008). Use of a recombinant MBD protein fused to the Fc fragment of the human IgG protein is called MCIp (Gebhard et al. 2006b). MIRA has also been coupled with microarray analysis, and in principle can be used with NGS (Rauch et al. 2008). Finally, the high-affinity MBD of MBD2 protein used in COMPARE-MS has also been used to generate methylated DNA libraries for genome-wide analysis with NGS and tiling microarrays (Serre et al. 2010; Yegnasubramanian et al. 2006).

These approaches are highly powerful because they generate a positive signal for methylated regions. When coupled with NGS or genome-wide tiling microarrays, they can allow truly unbiased analysis of genome-wide methylation patterns at a small fraction of the cost of carrying out sodium bisulfite shotgun sequencing. This is unlike restriction enzyme based approaches, which can also be cost effective, but only provide information at the recognition sites of the restriction enzymes. One major challenge in the analysis of these approaches is that the signal at a given location is dependent on at least two factors: (a) the extent of methylation at any given location; and (b) the density of methylation in a given fragment. For example, the same signal may be obtained for a region in which there is a relatively low 5meCpG density, but all of the input alleles are methylated as compared to a region where there is a very high 5meCpG density, but only a small fraction of input alleles are methylated. Therefore, although some analytical methods have been proposed to correct for this (Down et al. 2008), these approaches are not yet fully quantitative. This limitation is also true for the methylation-specific restriction enzyme (e.g. McrBc in CHARM and MethylScope) based approaches.

4.5 Strengths and Weaknesses of the Various Approaches for Genome-Wide DNA Methylation Analysis

The major advantage of bisulfite based genome-wide DNA methylation analyses is the ability to obtain nucleotide level resolution and quantitative measurement of methylation at each cytosine. However, despite this tremendous advantage, there are also several limitations. First, in order to achieve true complete genome coverage, methods such as bisulfite shotgun genomic sequencing must be used, which remain extremely costly and laborious despite precipitous declines in the cost of sequencing with the advent of NGS. Second, as discussed above, alignment of bisulfite converted genomic DNA can be challenging and improvements in these algorithms are needed. Third, biases can result from incomplete conversion of unmethylated cytosines and from preferential amplification of specific methylation configurations after conversion.

The major advantages of MSRE based approaches are that they are fairly cost-effective methods for enriching either the methylated or unmethylated portion of the genome. The major disadvantages include the limitation that only CpG sites interrogated by the restriction enzyme can be analyzed. When used with microarrays, algorithms to correct for probe effects and cross-hybridization biases must be used to improve signal to noise ratios. Additionally, biases can be introduced if there are mutations or polymorphisms that alter the recognition sequence, and these should be corrected for with appropriate controls (e.g. use of isoschizomers or control treatments with M.SssI to methylate all CpG sites). Also, since the processing of samples for use with microarrays or NGS often requires amplification, the different fragment sizes resulting from restriction enzyme digestion can produce biases in analysis.

The major strengths of the affinity based enrichment are that these approaches are highly cost-effective and capable of providing methylation across the entire genome without bias to specific sequences. These methods are currently the most promising of all the strategies for cost-effective, high-resolution, DNA methylation mapping for these reasons. Disadvantages include the fact that the signal is not only dependent on the extent of methyation across alleles, but also on the density of methylation within a given allele (this also holds true for the McrBc based restriction enzyme approaches since this enzyme preferentially cuts at regions with a higher density of mCpG). Another disadvantage is that, when analyzed with microarrays especially, there can be significant probe effects and GC content effects creating bias that must be accounted for in downstream analyses.

4.6 Detection of Methylated DNA by Physical Properties: The Horizon for Massively Parallel, Genome-Wide DNA Methylation Analysis

In the near horizon, improvements in single molecule sequencing may allow the direct detection of DNA methylation without the need for intermediate processing with sodium-bisulfite, restriction enzymes, or affinity enrichment. Pacific Biosciences, for instance, is introducing a third-generation sequencing platform that can monitor the incorporation of fluorescently labeled nucleotides by a DNA polymerase to a growing DNA strand in real time for rapid, and long read-length massively parallel sequencing (Eid et al. 2009). They propose that since they can detect the change in time the polymerase takes to incorporate a guanine opposite a methyl-cytosine as opposed to a cytosine, they may be able to detect DNA methylation directly (http://www.pacificbiosciences.com) (Flusberg et al. 2010). Other technologies, such as nanopore sequencing (Clarke et al. 2009), have also proposed that physical parameters, such as absorbance, conductance, electron resonance, and others, can be used to directly read sequence including the presence of 5meC instead of C, or even damaged DNA bases instead of native bases. Such innovations could usher a new era for integrated understanding of genetic and epigenetic alterations in cancer research.

References

Ball MP, Li JB, Gao Y, Lee JH, LeProust EM, Park IH, Xie B, Daley GQ, Church GM (2009) Targeted and genome-scale strategies reveal gene-body methylation signatures in human cells. Nat Biotechnol 27:361–368.

Bastian PJ, Palapattu GS, Lin X, Yegnasubramanian S, Mangold LA, Trock B, Eisenberger MA, Partin AW, Nelson WG (2005) Preoperative serum DNA GSTP1 CpG island hypermethylation and the risk of early prostate-specific antigen recurrence following radical prostatectomy. Clin Cancer Res 11:4037–4043.

Bedford MT, van Helden PD (1987) Hypomethylation of DNA in pathological conditions of the human prostate. Cancer Res 47:5274–5276.

Bibikova M, Fan JB (2009) GoldenGate assay for DNA methylation profiling. Methods Mol Biol 507:149–163.

Bibikova M, Le J, Barnes B, Saedinia-Melnyk S, Zhou L, Shen R, Gunderson KL (2009) Genome-wide DNA methylation profiling using Infinium assay. Epigenomics 1:177–200.

Bird AP (1986) CpG-rich islands and the function of DNA methylation. Nature 321:209–213.

Bird AP, Southern EM (1978) Use of restriction enzymes to study eukaryotic DNA methylation: I. The methylation pattern in ribosomal DNA from *Xenopus laevis*. J Mol Biol 118:27–47.

Bostick M, Kim JK, Esteve PO, Clark A, Pradhan S, Jacobsen SE (2007) UHRF1 plays a role in maintaining DNA methylation in mammalian cells. Science 317:1760–1764.

Brunner AL, Johnson DS, Kim SW, Valouev A, Reddy TE, Neff NF, Anton E, Medina C, Nguyen L, Chiao E, Oyolu CB, Schroth GP, Absher DM, Baker JC, Myers RM (2009) Distinct DNA methylation patterns characterize differentiated human embryonic stem cells and developing human fetal liver. Genome Res 19:1044–1056.

Cadieux B, Ching TT, Vandenberg SR, Costello JF (2006) Genome-wide hypomethylation in human glioblastomas associated with specific copy number alteration, methylenetetrahydrofolate reductase allele status, and increased proliferation. Cancer Res 66:8469–8476.

Challita PM, Skelton D, el-Khoueiry A, Yu XJ, Weinberg K, Kohn DB (1995) Multiple modifications in cis elements of the long terminal repeat of retroviral vectors lead to increased expression and decreased DNA methylation in embryonic carcinoma cells. J Virol 69:748–755.

Chapman V, Forrester L, Sanford J, Hastie N, Rossant J (1984) Cell lineage-specific undermethylation of mouse repetitive DNA. Nature 307:284–286.

Clarke J, Wu HC, Jayasinghe L, Patel A, Reid S, Bayley H (2009) Continuous base identification for single-molecule nanopore DNA sequencing. Nat Nanotechnol 4:265–270.

Costello JF, Fruhwald MC, Smiraglia DJ, Rush LJ, Robertson GP, Gao X, Wright FA, Feramisco JD, Peltomaki P, Lang JC, Schuller DE, Yu L, Bloomfield CD, Caligiuri MA, Yates A, Nishikawa R, Su Huang H, Petrelli NJ, Zhang X, O'Dorisio MS, Held WA, Cavenee WK, Plass C (2000) Aberrant CpG-island methylation has non-random and tumour-type-specific patterns. Nat Genet 24:132–138.

Cottrell SE, Distler J, Goodman NS, Mooney SH, Kluth A, Olek A, Schwope I, Tetzner R, Ziebarth H, Berlin K (2004) A real-time PCR assay for DNA-methylation using methylation-specific blockers. Nucleic Acids Res 32:e10.

Cross SH, Charlton JA, Nan X, Bird AP (1994) Purification of CpG islands using a methylated DNA binding column. Nat Genet 6:236–244.

Down TA, Rakyan VK, Turner DJ, Flicek P, Li H, Kulesha E, Graf S, Johnson N, Herrero J, Tomazou EM, Thorne NP, Backdahl L, Herberth M, Howe KL, Jackson DK, Miretti MM, Marioni JC, Birney E, Hubbard TJ, Durbin R, Tavare S, Beck S (2008) A Bayesian deconvolution strategy for immunoprecipitation-based DNA methylome analysis. Nat Biotechnol 26:779–785.

Dupont JM, Tost J, Jammes H, Gut IG (2004) De novo quantitative bisulfite sequencing using the pyrosequencing technology. Anal Biochem 333:119–127.

Eads CA, Danenberg KD, Kawakami K, Saltz LB, Blake C, Shibata D, Danenberg PV, Laird PW (2000) MethyLight: a high-throughput assay to measure DNA methylation. Nucleic Acids Res 28:E32.

Eckhardt F, Lewin J, Cortese R, Rakyan VK, Attwood J, Burger M, Burton J, Cox TV, Davies R, Down TA, Haefliger C, Horton R, Howe K, Jackson DK, Kunde J, Koenig C, Liddle J, Niblett D, Otto T, Pettett R, Seemann S, Thompson C, West T, Rogers J, Olek A, Berlin K, Beck S (2006) DNA methylation profiling of human chromosomes 6, 20 and 22. Nat Genet 38:1378–1385.

Ehrich M, Nelson MR, Stanssens P, Zabeau M, Liloglou T, Xinarianos G, Cantor CR, Field JK, van den Boom D (2005) Quantitative high-throughput analysis of DNA methylation patterns by base-specific cleavage and mass spectrometry. Proc Natl Acad Sci U S A 102:15785–15790.

Ehrlich M (2002) DNA methylation in cancer: too much, but also too little. Oncogene 21:5400–5413.

Eid J, Fehr A, Gray J, Luong K, Lyle J, Otto G, Peluso P, Rank D, Baybayan P, Bettman B, Bibillo A, Bjornson K, Chaudhuri B, Christians F, Cicero R, Clark S, Dalal R, Dewinter A, Dixon J, Foquet M, Gaertner A, Hardenbol P, Heiner C, Hester K, Holden D, Kearns G, Kong X, Kuse R, Lacroix Y, Lin S, Lundquist P, Ma C, Marks P, Maxham M, Murphy D, Park I, Pham T, Phillips M, Roy J, Sebra R, Shen G, Sorenson J, Tomaney A, Travers K, Trulson M, Vieceli J, Wegener J, Wu D, Yang A, Zaccarin D, Zhao P, Zhong F, Korlach J, Turner S (2009) Real-time DNA sequencing from single polymerase molecules. Science 323:133–138.

Esteller M, Corn PG, Baylin SB, Herman JG (2001) A gene hypermethylation profile of human cancer. Cancer Res 61:3225–3229.

Feinberg AP, Tycko B (2004) The history of cancer epigenetics. Nat Rev Cancer 4:143–153.

Feinberg AP, Vogelstein B (1983a) Hypomethylation distinguishes genes of some human cancers from their normal counterparts. Nature 301:89–92.

Feinberg AP, Vogelstein B (1983b) Hypomethylation of ras oncogenes in primary human cancers. Biochem Biophys Res Commun 111:47–54.

Feinberg AP, Gehrke CW, Kuo KC, Ehrlich M (1988) Reduced genomic 5-methylcytosine content in human colonic neoplasia. Cancer Res 48:1159–1161.

Feinberg AP, Cui H, Ohlsson R (2002) DNA methylation and genomic imprinting: insights from cancer into epigenetic mechanisms. Semin Cancer Biol 12:389–398.

Flusberg BA, Webster DR, Lee JH, Travers KJ, Olivares EC, Clark TA, Korlach J, Turner SW (2010) Direct detection of DNA methylation during single-molecule, real-time sequencing. Nat Methods 7:461–465.

Frigola J, Ribas M, Risques RA, Peinado MA (2002) Methylome profiling of cancer cells by amplification of inter-methylated sites (AIMS). Nucleic Acids Res 30:e28.

Frommer M, McDonald LE, Millar DS, Collis CM, Watt F, Grigg GW, Molloy PL, Paul CL (1992) A genomic sequencing protocol that yields a positive display of 5-methylcytosine residues in individual DNA strands. Proc Natl Acad Sci U S A 89:1827–1831.

Gama-Sosa MA, Slagel VA, Trewyn RW, Oxenhandler R, Kuo KC, Gehrke CW, Ehrlich M (1983) The 5-methylcytosine content of DNA from human tumors. Nucleic Acids Res 11:6883–6894.

Gebhard C, Schwarzfischer L, Pham TH, Andreesen R, Mackensen A, Rehli M (2006a) Rapid and sensitive detection of CpG-methylation using methyl-binding (MB)-PCR. Nucleic Acids Res 34:e82.

Gebhard C, Schwarzfischer L, Pham TH, Schilling E, Klug M, Andreesen R, Rehli M (2006b) Genome-wide profiling of CpG methylation identifies novel targets of aberrant hypermethylation in myeloid leukemia. Cancer Res 66:6118–6128.

Gitan RS, Shi H, Chen CM, Yan PS, Huang TH (2002) Methylation-specific oligonucleotide microarray: a new potential for high-throughput methylation analysis. Genome Res 12: 158–164.

Goelz SE, Vogelstein B, Hamilton SR, Feinberg AP (1985) Hypomethylation of DNA from benign and malignant human colon neoplasms. Science 228:187–190.

Gowher H, Jeltsch A (2001) Enzymatic properties of recombinant Dnmt3a DNA methyltransferase from mouse: the enzyme modifies DNA in a non-processive manner and also methylates non-CpG [correction of non-CpA] sites. J Mol Biol 309:1201–1208.

Gu H, Bock C, Mikkelsen TS, Jager N, Smith ZD, Tomazou E, Gnirke A, Lander ES, Meissner A (2010) Genome-scale DNA methylation mapping of clinical samples at single-nucleotide resolution. Nat Methods 7:133–136.

Hayashizaki Y, Hirotsune S, Okazaki Y, Hatada I, Shibata H, Kawai J, Hirose K, Watanabe S, Fushiki S, Wada S et al (1993) Restriction landmark genomic scanning method and its various applications. Electrophoresis 14:251–258.

Herman JG, Graff JR, Myohanen S, Nelkin BD, Baylin SB (1996) Methylation-specific PCR: a novel PCR assay for methylation status of CpG islands. Proc Natl Acad Sci U S A 93:9821–9826.

Hodges E, Smith AD, Kendall J, Xuan Z, Ravi K, Rooks M, Zhang MQ, Ye K, Bhattacharjee A, Brizuela L, McCombie WR, Wigler M, Hannon GJ, Hicks JB (2009) High definition profiling of mammalian DNA methylation by array capture and single molecule bisulfite sequencing. Genome Res 19:1593–1605.

Ibrahim AE, Thorne NP, Baird K, Barbosa-Morais NL, Tavare S, Collins VP, Wyllie AH, Arends MJ, Brenton JD (2006) MMASS: an optimized array-based method for assessing CpG island methylation. Nucleic Acids Res 34:e136.

Irizarry RA, Ladd-Acosta C, Carvalho B, Wu H, Brandenburg SA, Jeddeloh JA, Wen B, Feinberg AP (2008) Comprehensive high-throughput arrays for relative methylation (CHARM). Genome Res 18:780–790.

Jones PA, Laird PW (1999) Cancer epigenetics comes of age. Nat Genet 21:163–167.

Jones PA, Liang G (2009) Rethinking how DNA methylation patterns are maintained. Nat Rev Genet 10:805–811.

Jorgensen HF, Adie K, Chaubert P, Bird AP (2006) Engineering a high-affinity methyl-CpG-binding protein. Nucleic Acids Res 34:e96.

Kaminskas E, Farrell A, Abraham S, Baird A, Hsieh LS, Lee SL, Leighton JK, Patel H, Rahman A, Sridhara R, Wang YC, Pazdur R (2005) Approval summary: azacitidine for treatment of myelodysplastic syndrome subtypes. Clin Cancer Res 11:3604–3608.

Kantarjian H, Issa JP, Rosenfeld CS, Bennett JM, Albitar M, DiPersio J, Klimek V, Slack J, de Castro C, Ravandi F, Helmer R III, Shen L, Nimer SD, Leavitt R, Raza A, Saba H (2006) Decitabine improves patient outcomes in myelodysplastic syndromes: results of a phase III randomized study. Cancer 106:1794–1803.

Khulan B, Thompson RF, Ye K, Fazzari MJ, Suzuki M, Stasiek E, Figueroa ME, Glass JL, Chen Q, Montagna C, Hatchwell E, Selzer RR, Richmond TA, Green RD, Melnick A, Greally JM (2006) Comparative isoschizomer profiling of cytosine methylation: the HELP assay. Genome Res 16:1046–1055.

Klose RJ, Sarraf SA, Schmiedeberg L, McDermott SM, Stancheva I, Bird AP (2005) DNA binding selectivity of MeCP2 due to a requirement for A/T sequences adjacent to methyl-CpG. Mol Cell 19:667–678.

Laird PW (2003) The power and the promise of DNA methylation markers. Nat Rev Cancer 3:253–266.

Laird PW (2010) Principles and challenges of genome-wide DNA methylation analysis. Nat Rev Genet 11:191–203.

Lee WH, Morton RA, Epstein JI, Brooks JD, Campbell PA, Bova GS, Hsieh WS, Isaacs WB, Nelson WG (1994) Cytidine methylation of regulatory sequences near the pi-class glutathione S-transferase gene accompanies human prostatic carcinogenesis. Proc Natl Acad Sci U S A 91:11733–11737.

Li M, Chen WD, Papadopoulos N, Goodman SN, Bjerregaard NC, Laurberg S, Levin B, Juhl H, Arber N, Moinova H, Durkee K, Schmidt K, He Y, Diehl F, Velculescu VE, Zhou S, Diaz LA Jr, Kinzler KW, Markowitz SD, Vogelstein B (2009) Sensitive digital quantification of DNA methylation in clinical samples. Nat Biotechnol 27:858–863.

Liang G, Gonzalgo ML, Salem C, Jones PA (2002) Identification of DNA methylation differences during tumorigenesis by methylation-sensitive arbitrarily primed polymerase chain reaction. Methods 27:150–155.

Lister R, Pelizzola M, Dowen RH, Hawkins RD, Hon G, Tonti-Filippini J, Nery JR, Lee L, Ye Z, Ngo QM, Edsall L, Antosiewicz-Bourget J, Stewart R, Ruotti V, Millar AH, Thomson JA, Ren B, Ecker JR (2009) Human DNA methylomes at base resolution show widespread epigenomic differences. Nature 462:315–322.

Mann BS, Johnson JR, Cohen MH, Justice R, Pazdur R (2007) FDA approval summary: vorinostat for treatment of advanced primary cutaneous T-cell lymphoma. Oncologist 12:1247–1252.

Meissner A, Gnirke A, Bell GW, Ramsahoye B, Lander ES, Jaenisch R (2005) Reduced representation bisulfite sequencing for comparative high-resolution DNA methylation analysis. Nucleic Acids Res 33:5868–5877.

Meissner A, Mikkelsen TS, Gu H, Wernig M, Hanna J, Sivachenko A, Zhang X, Bernstein BE, Nusbaum C, Jaffe DB, Gnirke A, Jaenisch R, Lander ES (2008) Genome-scale DNA methylation maps of pluripotent and differentiated cells. Nature 454:766–770.

Norris DP, Brockdorff N, Rastan S (1991) Methylation status of CpG-rich islands on active and inactive mouse X chromosomes. Mamm Genome 1:78–83.

Oda M, Glass JL, Thompson RF, Mo Y, Olivier EN, Figueroa ME, Selzer RR, Richmond TA, Zhang X, Dannenberg L, Green RD, Melnick A, Hatchwell E, Bouhassira EE, Verma A, Suzuki M, Greally JM (2009) High-resolution genome-wide cytosine methylation profiling with simultaneous copy number analysis and optimization for limited cell numbers. Nucleic Acids Res 37:3829–3839.

Onyango P, Jiang S, Uejima H, Shamblott MJ, Gearhart JD, Cui H, Feinberg AP (2002) Monoallelic expression and methylation of imprinted genes in human and mouse embryonic germ cell lineages. Proc Natl Acad Sci U S A 99:10599–10604.

Ordway JM, Bedell JA, Citek RW, Nunberg A, Garrido A, Kendall R, Stevens JR, Cao D, Doerge RW, Korshunova Y, Holemon H, McPherson JD, Lakey N, Leon J, Martienssen RA, Jeddeloh JA (2006) Comprehensive DNA methylation profiling in a human cancer genome identifies novel epigenetic targets. Carcinogenesis 27:2409–2423.

Plass C, Shibata H, Kalcheva I, Mullins L, Kotelevtseva N, Mullins J, Kato R, Sasaki H, Hirotsune S, Okazaki Y, Held WA, Hayashizaki Y, Chapman VM (1996) Identification of Grf1 on mouse chromosome 9 as an imprinted gene by RLGS-M. Nat Genet 14:106–109.

Pollack Y, Stein R, Razin A, Cedar H (1980) Methylation of foreign DNA sequences in eukaryotic cells. Proc Natl Acad Sci U S A 77:6463–6467.

Ramsahoye BH, Biniszkiewicz D, Lyko F, Clark V, Bird AP, Jaenisch R (2000) Non-CpG methylation is prevalent in embryonic stem cells and may be mediated by DNA methyltransferase 3a. Proc Natl Acad Sci U S A 97:5237–5242.

Rauch T, Pfeifer GP (2005) Methylated-CpG island recovery assay: a new technique for the rapid detection of methylated-CpG islands in cancer. Lab Invest 85:1172–1180.

Rauch TA, Zhong X, Wu X, Wang M, Kernstine KH, Wang Z, Riggs AD, Pfeifer GP (2008) High-resolution mapping of DNA hypermethylation and hypomethylation in lung cancer. Proc Natl Acad Sci U S A 105:252–257.

Razin A, Riggs AD (1980) DNA methylation and gene function. Science 210:604–610.

Rodriguez J, Frigola J, Vendrell E, Risques RA, Fraga MF, Morales C, Moreno V, Esteller M, Capella G, Ribas M, Peinado MA (2006) Chromosomal instability correlates with genome-wide DNA demethylation in human primary colorectal cancers. Cancer Res 66:8462–9468.

Serre D, Lee BH, Ting AH (2010) MBD-isolated genome sequencing provides a high-throughput and comprehensive survey of DNA methylation in the human genome. Nucleic Acids Res 38:391–399.

Sharif J, Muto M, Takebayashi S, Suetake I, Iwamatsu A, Endo TA, Shinga J, Mizutani-Koseki Y, Toyoda T, Okamura K, Tajima S, Mitsuya K, Okano M, Koseki H (2007) The SRA protein Np95 mediates epigenetic inheritance by recruiting Dnmt1 to methylated DNA. Nature 450:908–912.

Shinar D, Yoffe O, Shani M, Yaffe D (1989) Regulated expression of muscle-specific genes introduced into mouse embryonal stem cells: inverse correlation with DNA methylation. Differentiation 41:116–126.

Sidransky D (2002) Emerging molecular markers of cancer. Nat Rev Cancer 2:210–219.

Siegfried Z, Cedar H (1997) DNA methylation: a molecular lock. Curr Biol 7:R305–307.

Siegfried Z, Eden S, Mendelsohn M, Feng X, Tsuberi BZ, Cedar H (1999) DNA methylation represses transcription in vivo. Nat Genet 22:203–206.

Singer J, Roberts-Ems J, Riggs AD (1979) Methylation of mouse liver DNA studied by means of the restriction enzymes msp I and hpa II. Science 203:1019–1021.

Singer-Sam J, Grant M, LeBon JM, Okuyama K, Chapman V, Monk M, Riggs AD (1990a) Use of a HpaII-polymerase chain reaction assay to study DNA methylation in the Pgk-1 CpG island of mouse embryos at the time of X-chromosome inactivation. Mol Cell Biol 10:4987–4989.

Singer-Sam J, LeBon JM, Tanguay RL, Riggs AD (1990b) A quantitative HpaII-PCR assay to measure methylation of DNA from a small number of cells. Nucleic Acids Res 18:687.

Song F, Smith JF, Kimura MT, Morrow AD, Matsuyama T, Nagase H, Held WA (2005) Association of tissue-specific differentially methylated regions (TDMs) with differential gene expression. Proc Natl Acad Sci U S A 102:3336–3341.

Suzuki K, Suzuki I, Leodolter A, Alonso S, Horiuchi S, Yamashita K, Perucho M (2006) Global DNA demethylation in gastrointestinal cancer is age dependent and precedes genomic damage. Cancer Cell 9:199–207.

Taylor KH, Kramer RS, Davis JW, Guo J, Duff DJ, Xu D, Caldwell CW, Shi H (2007) Ultradeep bisulfite sequencing analysis of DNA methylation patterns in multiple gene promoters by 454 sequencing. Cancer Res 67:8511–8518.

The Cancer Genome Atlas Project (2008) Comprehensive genomic characterization defines human glioblastoma genes and core pathways. Nature 455:1061–1068.

Thomassin H, Kress C, Grange T (2004) MethylQuant: a sensitive method for quantifying methylation of specific cytosines within the genome. Nucleic Acids Res 32:e168.

Tilghman SM (1999) The sins of the fathers and mothers: genomic imprinting in mammalian development. Cell 96:185–193.

Tolberg ME, Funderburk SJ, Klisak I, Smith SS (1987) Structural organization and DNA methylation patterning within the mouse L1 family. J Biol Chem 262:11167–11175.

Wang RY, Gehrke CW, Ehrlich M (1980) Comparison of bisulfite modification of 5-methyldeoxycytidine and deoxycytidine residues. Nucleic Acids Res 8:4777–4790.

Weber M, Hellmann I, Stadler MB, Ramos L, Paabo S, Rebhan M, Schubeler D (2007) Distribution, silencing potential and evolutionary impact of promoter DNA methylation in the human genome. Nat Genet 39:457–466.

Yan PS, Potter D, Deatherage DE, Huang TH, Lin S (2009) Differential methylation hybridization: profiling DNA methylation with a high-density CpG island microarray. Methods Mol Biol 507:89–106.

Yegnasubramanian S, Kowalski J, Gonzalgo ML, Zahurak M, Piantadosi S, Walsh PC, Bova GS, De Marzo AM, Isaacs WB, Nelson WG (2004) Hypermethylation of CpG islands in primary and metastatic human prostate cancer. Cancer Res 64:1975–1986.

Yegnasubramanian S, Lin X, Haffner MC, DeMarzo AM, Nelson WG (2006) Combination of methylated-DNA precipitation and methylation-sensitive restriction enzymes (COMPARE-MS) for the rapid, sensitive and quantitative detection of DNA methylation. Nucleic Acids Res 34:e19.

Zhang X, Yazaki J, Sundaresan A, Cokus S, Chan SW, Chen H, Henderson IR, Shinn P, Pellegrini M, Jacobsen SE, Ecker JR (2006) Genome-wide high-resolution mapping and functional analysis of DNA methylation in arabidopsis. Cell 126:1189–1201.

Chapter 5
Use of Expression Microarrays in Cancer Research

Jun Luo and Yidong Chen

Abstract Since its inception more than 15 years ago, the rapidly evolving array-based gene expression technology has been widely adopted and become an indispensable tool in cancer research. In this chapter, we will discuss the various platforms and the corresponding technical and analytical steps including study design, sample selection and processing, data generation and data analysis. We will identify and discuss key issues that may affect the reliability and precision of end-point array results, as well as common pitfalls that influence the interpretation of the analytical results.

5.1 Overall Goal of Expression Microrarray Analysis in Cancer Research

While recognizing the common hallmarks of human cancer shared among the more than 100 different human cancers (Hanahan and Weinberg 2000), it is equally important to appreciate and analyze the diversity in the natural history and clinical behavior among tumors arising from different organ sites or even tumors originating from the same organ site. Human cancers can develop through exposure to different etiological factors and accumulation of different molecular alterations en route to malignancies associated with variable clinical outcome. It is well established that the initiation and progression of human cancer is accompanied by a myriad of DNA-level alterations in each stage of the multi-step process (Nelson et al. 2003; Carbone and Pass 2004). The interactions between these alterations and the tumor microenvironment shape the individual phenotype of human cancer. Therefore, a central theme in cancer research has been the delineation of molecular changes contributing

J. Luo (✉)
Department of Urology, Johns Hopkins University School of Medicine, Baltimore, MD, USA
e-mail: jluo1@jhmi.edu

S. Yegnasubramanian and W.B. Isaacs (eds.), *Modern Molecular Biology*, Applied Bioinformatics and Biostatistics in Cancer Research, DOI 10.1007/978-0-387-69745-1_5, © Springer Science+Business Media, LLC 2010

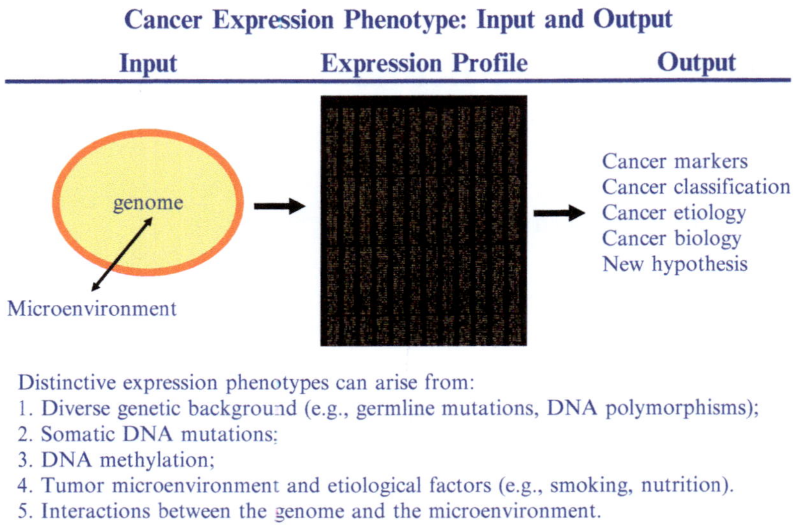

Fig. 5.1 The expression phenotype: input and output. Distinctive expression phenotypes can arise from (1) diverse genetic backgrounds (e.g., germline mutations, DNA polymorphisms); (2) somatic DNA mutations; (3) DNA methylation; (4) tumor microenvironment and etiological factors (e.g., smoking, nutrition); (5) interactions between the genome and the microenvironment

to each stage of cancer development and each cancer subtype, with the ultimate goal of translating them into clinical tools for individualized risk assessment, detection, prognosis, and targeted therapy. Array-based expression analysis tools measure the global profile of expressed gene products, which reflects the combined phenotypic output of cumulative variations in genetic and epigenetic profile as well as the tumor microenvironment, and aim to develop a surrogate for the corresponding cellular physiological state and the clinical behavior of human cancers (Fig. 5.1). Since both genomic and environmental factors, either naturally occurring or experimentally controlled, have the potential to disrupt the multi-level interactions between genes and gene products, the resulted phenotypic output assessed by genome-wide expression profiling also provides functional insight and help to generate new hypothesis for detailed mechanistic investigation in experimental models (Fig. 5.1). Therefore, array-based expression analysis has been established as one of the most important approaches for molecular studies of human cancer.

5.2 An Overview of Major Components of an Expression Microarray Study

For any given microarray study, researchers often face numerous options ranging from the choice of array platforms and various technical approaches in the data generation phase of the study, to the choice of analytical approaches once the

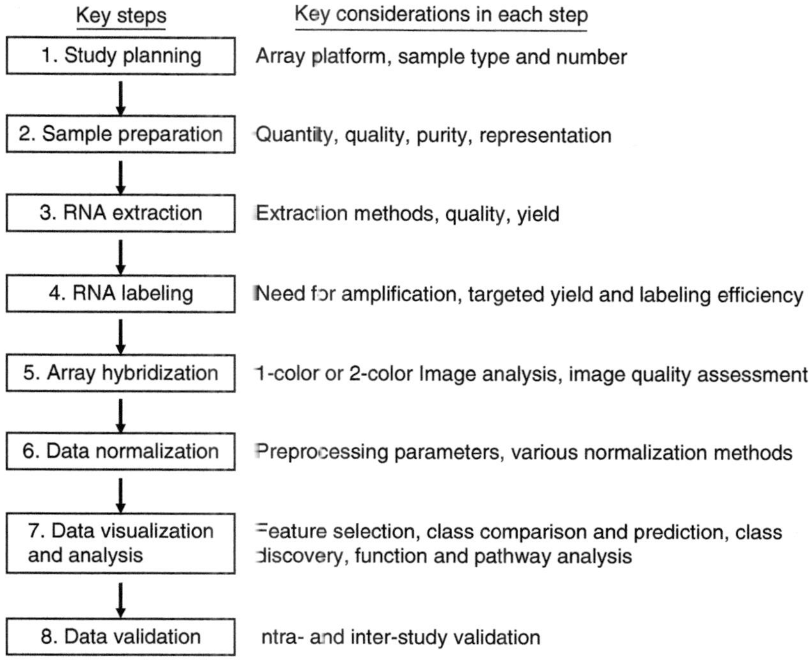

Fig. 5.2 Overview of experimental steps and considerations in microarray studies

data is generated. While commercial products and services often provide specific recommendations and quality controlled technical procedures tailored to each array platform, each of the key steps of a microarray project (Fig. 5.2) can be performed with flexibility to suit various practical considerations intrinsic to each individual study. A deep understanding of the basis for each specific option is required of the study investigators at the planning stage for study design, particularly for large studies involving many patient samples. The most important decision made at the planning stage pertains to the array platform that would maximize the quality of the molecular data during the project period, which is often tied to the size of the study as well as the available platform specific data generation and data analysis approaches. Rigorous quality control measures should be implemented in each of data generation steps (steps 2–5, Fig. 5.2) to facilitate the downstream analytical and validation steps (steps 6–8, Fig. 5.2) that would lead to study conclusions.

5.3 Diversity of Array Platforms

Expression microarrays consist of ordered, addressable arrays of microscopic DNA probes manufactured on a solid support. The DNA probes (also named features or array elements) detect the presence and relative abundance of gene transcripts (also named targets), often labeled, through complementary hybridization. The prototype of such arrays was first reported by Southern et al. (1992), and followed by a number of platforms that laid the foundation for advanced microarray technologies commonly used today (Pease et al. 1994; Schena et al. 1995; Walt 2000; Hughes et al. 2001). Commercial expression arrays are by far the most developed among high throughput molecular profiling devices that measure molecular abundance at the whole-genome level. Currently, over a dozen commercial vendors offer a diverse range of expression microarray products and services that differ in the type of array platforms and corresponding labeling strategies, hardware requirements (e.g., array scanner), and analytical software compatibilities. Much of the differences among these products and a number of in-house varieties have been previously described (Luo et al. 2003; Hardiman 2004; Ahmed 2006; Elvidge 2006). Although in-house spotted cDNA microarrays were once the most prevalent platform, and will likely continue to play a significant role in academic research, many researchers shifted to the more competitive commercial platforms due to their lowered cost, much improved quality, standardization, and availability of supporting services. Table 5.1 lists the number of microarray datasets generated using different array platforms that were submitted to NCBI Gene Expression Omnibus (as of August 2, 2008), a major portal for deposition and retrieval of microarray data (Edgar et al. 2002; Barrett and Edgar 2006). For the sole purpose of illustrating the wide range of platform specific differences and their implications in the various steps of expression microarray analysis, we will focus our discussion on two vastly different high-density oligonucleotide expression microarrays that currently dominate the commercial expression array market, Affymetrix GeneChip and Agilent 60-mer expression microarrays. The two other relatively popular platforms, Illumina and Nimblegen, will be briefly discussed in the appropriate context.

Affymetrix has been the leader in utilizing photolithography, a photo-masking technology from the semiconductor industry, for light-directed in situ synthesis of DNA probes of 25 bp in size on a silicon wafer (Pease et al. 1994). Step-wise synthesis efficiency for the current technology is approximately 95%, limiting the yield of full-length probes. This relatively lower synthesis efficiency and limited probe length (25 bp) has led to measurement imprecision and low sensitivity of signal detection that are compensated by the inclusion of probes sets (multiple probes for each target transcript) and mismatch (MM) probes paired with the perfect match (PM) probes. The MM probes differ from the PM probes by one nucleotide at the

Table 5.1 Microarray data sets in GEO

Platform	RNA	DNA	Others
Affymetrix	4,136	158	17
Agilent	451	88	6
Illumina	107	15	2
Nimblegen	74	122	1
Other platforms	3,512	389	49
Total	8,280	772	75

central position. PM minus MM was an early method of removing background hybridization signal, but later replaced with more popular normalization methods that actually discard the MM data (Bolstad et al. 2003). The current Human U133 Plus 2.0 version Affymetrix GeneChip contains ten probe pairs (i.e., 20 features) for each of the 47,000 target transcripts, with each feature sized at 11 μm. The published detection sensitivity for the current product line is 1:100,000. As manufacturing and detection technology advances, these GeneChip arrays will accommodate more features (up to five million features) and the corresponding feature size will continue to shrink (down to 5 μm/feature). The Affymetrix platform employs biotin labeling and streptavidin–phycoerythrin fluorescence detection using a platform specific scanner, and only allows one labeled sample for hybridization to the arrayed probes.

Agilent oligonucleotide expression microarrays were manufactured using the ink-jet technology (Hughes et al. 2001), which involves delivery of nucleosides for solution-based in situ synthesis of 60-mer oligonucleotides on the surface of glass slides. The synthesis is based on the phosphoramidite chemistry with step-wise efficiency approaching 99.5% and high yield of full-length 60-mer probes (~75%). The currently published detection sensitivity of the Agilent arrays is 1:300,000. The enhanced detection sensitivity is largely a function of the probe length and purity. The current product line includes the human whole genome expression microarray with four arrays of 44 K probes, allowing simultaneous assays of four test samples in a single slide. Most target transcripts are detected by a single probe, designed based on the three prime sequence of the target transcript. The feature size is approximately 70 μm, leaving room for further reduction of the feature size to accommodate more probes for higher density expression arrays such as exon arrays. The Agilent platform is compatible with both single- and dual-color analysis. In dual-color analysis, two samples are labeled with different fluorescent dyes (Cy-5 and Cy-3) and cohybridized with the arrayed probes.

5.4 Considerations Related to Platform Choices

The availability of a wide assortment of array platforms has motivated many detailed studies to compare the measurement precision and accuracy of the different platforms. Earlier on, cross-platform comparisons and integration of gene expression results often found generally low concordance among different platforms and study sites, yet multiple sources of variation have been identified that can be effectively controlled to improve cross-platform concordance (Bammler et al. 2005; Irizarry et al. 2005). Sources of cross-platform variation include the probe content and probe annotation, data generation methods, data preprocessing and normalization methods. Higher concordance can be achieved following protocol standardization and the use of the same commercial platform (Bammler et al. 2005; Dobbin et al. 2005a). Much improved concordance reported in more recent studies from the MicroArray Quality Control Consortium (MAQC) (Kuo et al. 2006;

Shi et al. 2006) reflected the incorporation of knowledge gained from previous studies as well as the evolving array technologies with improved measurement precision. However, there are still issues related to the discordance among the different platforms, and between array data and qRT-PCR validation data. Much of the discordance can be attributed to the differences of the array content (i.e., probes). All commercial vendors can produce excellent correlation between their array data and qRT-PCR data, however these data are often generated using the same probe sequences for both platforms. It is anticipated that continued improvement of the probe content in commercial arrays will lead to a further decline in cross-platform variation.

The choice of which microarray platform to use, therefore, should be carefully considered in order to obtain array data that is not only of high precision (i.e., reproducible under similar conditions) but also of high accuracy (i.e., reproducible under different conditions). The nature of the particular study should be the principle guide for determining the most suitable array platform. Spotted glass cDNA microarrays have been prevalently used in academic labs and relevant studies contributed to a rich collection of the current literature (Table 5.1). Following advances in genome sequencing and annotation as well as the lowered cost and enhanced custom compatibility of commercial oligonucleotide arrays, however, we anticipate declining popularity of this platform in the years to come mainly due to limitations on density as well as systemic variations that are difficult to manage and standardize. The choice among oligonucleotide based platforms depends on many factors such as desired probe content, whether two-color design is needed, RNA quantity and quality, technical sophistication required for data generation and analysis, and overall cost. Commercial expression arrays have fallen into two broad categories: short 25-mer arrays (e.g., Affymetrix GeneChip® array) and long 50-mer to 70-mer oligo arrays (e.g., Agilent, NimbleGen, Illumina). Much of the claimed advantages for each specific platform lack solid data support and the debate is still unsettled regarding whether 25-mer probes provides sufficient thermodynamic stability to prevent mishybridization, and whether longer oligonucleotide probes are sufficiently specific to distinguish highly homologous genes. For array platforms with fixed probe lengths (e.g., Affymetrix, Agilent), measurement precision can be affected by thermodynamic biases introduced by variations of T_m of the probes. This bias can be corrected by improved probe design as well as probe length adjustments to generate isothermal probe sets (e.g., NimbleGen).

5.5 Sample Type and Size

Multiple testing is generic to any microarray study, requiring adjustment of standard p values to reduce false positive results. The most popular approach for controlling multiple testing is the calculation of false discovery rate (FDR) (Reiner et al. 2003), which is the expected proportion of false discoveries among the significant events defined by a threshold value. Well-designed microarray studies will have high power to detect expression differences of desired magnitude while minimizing the FDR. Therefore study design should take into consideration of the variation among the sample types and the number of samples required to achieve defined statistical power, so that the likelihood of meeting the study goals can be assessed to guide interpretation of study results as well as the scale of follow-up validation studies. Study power and required sample size depends on the variance of individual measurements contributed by both biological variation and experimental measurement variation. The relation between sample type and required samples size is often overlooked. With variation in technical measurement precision controlled and minimized by implementing quality control measures (in steps 2–5, Fig. 5.2), the required sample size will largely depend on properties intrinsic to the study subjects, the sampling and processing of which can create variations of vastly different magnitude. For example, many clinical cancer specimens are highly heterogeneous both within each tumor and among the different tumors. Variations in the tissue composition, surgical methods, processing time etc. all have the potential to introduce biological variations that are not the focus of the study. Therefore an estimation of the variance, often based on previous studies and pilot studies, would only be relevant if the sampling and tissue processing strategies are representative of the current study. Sample size calculation methods tailored to a variety of platforms, different experimental designs, and varying end-point analytical goals (e.g., class comparison, class prediction) have been described in detail along with the relevant online software (Yang et al. 2003; Dobbin and Simon 2005). Accurate estimation of variance and expected effect size remain a challenge in power analysis for microarray studies, yet incorporating sample size calculation into large-scale microarray experiments will be more feasible as the experimental procedures for expression microarray analysis are further refined and reliability of expression data further improved.

5.6 Generation of Expression Microarray Data

Biological samples used for a microarray study can be any type from which RNA of sufficient quality and quantity can be extracted. Cell lines treated under different conditions often provide a controlled environment for extraction of high quality RNA. However many studies will involve clinical specimens with varying degree of RNA degradation and intra-tissue heterogeneity. In this case quality control measures must be in place to track the tissue properties such as target tissue purity and integrity. The presence of infiltrating lymphocytes, other non-target lesions in the tissue specimen, the degree of autolysis, and the surgical source of the tissue specimen can all drastically alter expression profile and downstream analysis. In addition, while the most suitable tissues for array analysis are those processed by fresh freezing, expression profiling is increasingly been conducted with formalin fixation paraffin embedding (FFPE) specimens, which are often associated with poor RNA yield and inferior RNA quality. The two tissue sources will require entirely different methods for RNA extraction and many of the downstream steps (von Ahlfen et al. 2007). Following identification and isolation of the target tissues, total RNA can be extracted using phenol/chloroform or column-based purification methods and subjected to quantification and quality assessment, often using the Agilent Bioanalyzer or Nanodrop spectrophotometer. More often than not, certain RNA quality measures are implemented to ensure the exclusion of samples with inferior RNA quality from the study (Thompson et al. 2007; Weis et al. 2007). The method of choice for RNA labeling is influenced by the amount of RNA available and other parameters specific to the microarray platform. In many microarray applications, it is necessary to amplify RNA due to limited RNA quantity. Routinely, the "Eberwine" method of linear RNA amplification (Van Gelder et al. 1990) is integrated in the labeling protocols of many array platforms. In this method, cDNA synthesis is primed with an oligo-dT primer anchored by a core T7 DNA-dependent RNA polymerase binding sequence (T7 promoter). Following the generation of the template double stranded cDNA, the in vitro transcription (IVT) reaction catalyzed by the T7 polymerase can linearly amplify the mRNA, which constitutes approximately 1–5% of the total RNA, by up to 1,000-fold in one round of amplification. During the IVT reaction modified nucleotides are incorporated into the antisense RNA end-product. The modified nucleotides can be dye-labeled nucleotides that allow direct labeling, or modified nucleotides tagged by biotin or amino-allyl that are later linked to dye molecules compatible with detection with platform specific laser scanners. The bias introduced by a single round of linear amplification was minimal even when assessed well before high-quality commercial oligo arrays were available (Wang et al. 2000; Baugh et al. 2001), but can still be substantial if two rounds of amplification are performed today (Boelens et al. 2007). However, two rounds of amplification is often necessary to yield enough labeled products from laser-captured pure tissue lesions or degraded RNA samples such as those from formalin fixed and paraffin embedded clinical specimens. The bias introduced by linear amplification nevertheless is almost always outweighed by the benefits when the quantity of RNA is a limiting factor. In addition, the linear amplification method is preferred over PCR-based exponential amplification methods, but it is generally not desirable to go beyond two rounds of amplification. Exponential amplification techniques are not widely used but may show promise upon further refinement (Elvidge 2006). Linear amplification is biased toward the 3′ end of the transcripts and therefore may require corresponding array platforms with 3′ biased design. With two-round RNA amplification, highly reliable expression data can be readily derived from a few hundred cells provided that RNA (10 pg total RNA per cell) is not degraded.

Spotted glass cDNA arrays and some commercial long oligo arrays (e.g., Agilent) are permissive for the two color design in which RNA derived from two biological specimens can be labeled with different dyes (Cy-5 vs. Cy-3) and cohybridized onto the same array, while the Affymetrix platform only permits one labeled sample per array. This difference accounts for corresponding differences in array data normalization and analytical approaches for these different platforms. The main purpose of the two-color design is to control inter-array measurement variation, which has been minimized in the commercial arrays in use today (Patterson et al. 2006). However, the two-color design is still widely used not only for two-sample comparison but also mainly for comparison of gene expression patterns among samples measured against the common reference standard. One potential drawback of the two-color design is dye bias. The differential dye incorporation rate and energy output is considered the root cause for this dye bias (Cox et al. 2004). Dye bias often presents as an intensity-dependent deviation from perfect correlation between the two channels. Although the two-color design is the most accurate way to measure the relative differences between two samples because all conditions are identical other than the target transcripts, uncontrolled dye bias can completely eliminate this benefit. In our experience, dye bias can be effectively controlled by rigorous quality control and optimization of technical steps involved in microarray analysis (steps 2–5, Fig. 5.2). Dye bias can also be minimized by dye-swap replication (Rosenzweig et al. 2004), by preprocessing of array data to eliminate affected probes (typically those associated with low intensities), or by LOESS normalization (Dobbin et al. 2005b). Currently the Agilent expression array products use LOESS and a proprietary error model within the Feature Extraction software to produce a robust set of ratio measurements (see below) for experiments using the two-color design. The decision in regard to whether to use two-color design and how to control dye bias often falls into individual preferences.

Labeled samples (cDNA or cRNA) are hybridized onto the arrays with complementary probes by overnight incubation at platform specific temperatures, and the microarrays are subsequently processed for fluorescence detection using platform specific laser scanners that also have evolved in their resolution, throughput, and analytical tools to meet to demand of ever increasing array density. Much of the platform specific information as well as cross-platform compatibility can be found from the vendors' web sites.

5.7 Microarray Data Normalization

Various technical steps involved in an array experiments can generate technical biases that must be corrected for by normalization prior to detailed comparative analysis (Quackenbush 2002). Technical variation and measurement imprecision can originate from differences in dye incorporation and differential energy output of the dyes (dye bias), scanner variation (scanner bias), local or systemic background fluorescence (background bias), differences in probe design and related sensitivity and specificity of target detection (probe bias), differential RNA quality and quantity, and differential RNA degradation. A set of normalization steps and tools have been developed for expression profiling (Quackenbush 2002; Bolstad et al. 2003; Harr and Schlotterer 2006). Normalization can be applied to signal intensities or ratios measurements from the two-color platforms (Attoor et al. 2004). Ratios have been shown to be very precise and accurate, in the absence of dye bias, because of the kinetics of competitive hybridization and the controlled inter-array variation when measuring two mRNA species at the same time in the same hybridization (e.g., Agilent). Nevertheless, measurement of expression intensities from single-color data have become a common practice for studies involving a large set of samples. This is partly owing to the improved array platforms (Affymetrix). Currently the Agilent expression system (Wolber et al. 2006) uses a default LOESS normalization scheme coupled with an error model within Feature Extraction software without background correction, while the MAS5-normalization scheme (with present, marginal and absent flags) is the most commonly reported data normalization for the Affymetrix platform (Wu and Irizarry 2004). Robust Multiarray Averaging (RMA) is another method to normalize 1-color array data, with its derivative GC-RMA developed to account for probe GC content (Fan et al. 2005). It is worth noting that most normalization methods were developed at a time when tremendous technical biases and measurement imprecision were routine. Technical improvements in array manufacturing and standardization of technical steps have contributed to much improved accuracy and precision of microarray data. Normalization is no longer a demanding task in most expression microarray studies. However, cross-array discordance mainly attributable to the different probe content is still a major source of variation and may require probe-level normalization for some studies. Further probe optimization as well as in-depth knowledge on transcript splicing will lead to new versions of commercial oligo arrays that provide better concordance between the different platforms (Lee et al. 2007).

5.8 Expression Microarray Data Analysis

In a microarray experiment, the arrayed probes are often in abundant excess relative to the transcript targets. Therefore the signal intensities detected are generally proportional to the relative abundance of the transcript targets, provided that they do not saturate and are within the dynamic range of detection. Therefore for practical purposes signal intensities are often suggestive of the absolute abundance of the targets and can be used to guide the follow-up studies. In an analytical setting, however, inferences can only be made in the comparison of a gene in different samples but not in the comparison of expression abundance of one transcript versus another in the same sample, largely due to detection variations intrinsic to the individual probes. For a simple two-sample comparison and studies aimed at finding the most differentially expressed genes (e.g., the top genes induced by androgen treatment in one cell line at a fixed time point and fixed concentration), genes and target transcripts can be ranked by fold expression changes, or by T-statistic to account for technical measurement variation from the scanner – note that biological variation can not be addressed in this two-sample comparison example. Technical replicates, repeat assays of the same RNA samples, are often used to assess the reproducibility of the array platform. With rare exceptions such as the need to test the effect of various technical steps (e.g., the variation introduced by RNA amplification), technical replicates are rarely necessary in the design of microarray studies due to high degree of reproducibility of the commercial platforms under standardized conditions.

Class comparison, in which different categories of study samples are compared, is the objective of many expression microarray studies. It differs from the above two-sample comparison by inclusion of multiple biological replicates (i.e., multiple different samples per defined category). For example, transcripts that differentiate benign growth and malignant growth in the human prostate can be identified by comparing tissue specimens from many different surgical cases (Luo et al. 2001). In class comparison with multiple biological replicates (e.g., multiple patients) in each class, effective methods (Jeffery et al. 2006) exist for identification of differentially expressed genes while controlling false positive findings. False positive findings are inherent to microarray studies due to the large number of variables (probes) and small number of biological replicates. More often than not, these methods involve the use of modified versions of the T-statistic (Luo et al. 2001; Tusher et al. 2001) for gene ranking, and the use of permutation based methods to determine the cut-off values based on desired false discovery rates. Despite its inadequacy in statistical analysis (Smyth et al. 2003), fold-change based gene selection is still widely and legitimately used especially when biomarker discovery and validation is the main objective. For example, the "volcano plot" (Chen et al. 2007) can be used to select genes based on fold change in combination with less stringent *p* values. Class comparison studies involving multiple classes can be analyzed using modified F statistics or analysis of variance (ANOVA) (Jin et al. 2001). For comparisons involving different time points in multiple categories of samples, genes that change with time or fit a specified pattern of change can be identified (Storey et al. 2005). Univariate proportional hazards regression analysis (Bovelstad et al. 2007) can be used to identify genes that correlate with treatment outcome, which is often presented as survival time until a specified progression end-point. Although the statistical methods may vary depending on the expected magnitude of expression change as well as the type and number of sample categories, the common goal of all class comparison microarray studies is to define cut-off values to derive a set of differentially expressed genes while controlling the proportion of false discovery events. Inevitably, a subset of truly differentially expressed genes that do not pass the cut-off threshold will be declared as nonsignificant, leading to another measure termed "miss rate" that can also be calculated using similar approaches (Taylor et al. 2005). Genes demonstrating only subtle expression changes are particularly susceptible to high miss rates. This problem can be partially alleviated by analyzing grouped gene sets instead of individual genes. One example of this approach is the Gene Set Enrichment Analysis (GSEA) (Subramanian et al. 2005), developed to identify groups of coordinately regulated genes that could otherwise be missed using standard approaches. For gene set analysis, because the inter-relationships among the genes were defined a priori and thus used prospectively in the identification of gene sets, the GSEA approach is not as constrained by the statistical stringency required in identifying the individual genes. A few advanced versions of the approach have been developed for assessing whether the correlation between the gene sets and the sample categories is statistically significant (Tian et al. 2005; Kong et al. 2006; Jiang and Gentleman 2007).

Class prediction studies represent advanced investigations directly aimed at assessing the utility of expression data in predicting the class membership. Class prediction involves first the development of a class predictor (i.e., a prediction or classification model) based on a subset of samples, followed by performance assessment in a new set of samples, with the ultimate objective to predict the diagnostic, prognostic or drug response category. The prediction model consists of a mathematical function of gene vectors developed from class comparison and model fitting (Radmacher et al. 2002). Performance assessment involves assessment of prediction error rates in a new set of samples. Two basic strategies, split-sample and cross validation, have been widely used in class prediction studies. In the split-sample example, the classifier is built from a subset of the data (e.g., from two-thirds of the samples), named the training set, and used to estimate the prediction error rate in the remaining data, the test set (e.g., one third of the samples). This approach requires that the test set data must be completely independent of the training set. In other words, the test data should not be used in any way until a single model is developed from the model building process. This rule is commonly violated in a variety of forms in the published literature, leading to the common problem of model overfitting (Ransohoff 2004). The propensity of model overfitting in microarray studies can be exemplified by a near-perfect fit of the randomly and artificially generated training data (Simon et al. 2003). This split sample approach is not optimal for use when the number of samples is insufficient for meaningful splitting. The alternative cross validation method involves an iterative process of partitioning data into a large training set, and a validation data set consisting of only one or a few samples. In each iteration, a model is built from the training set and used to predict the class membership of the left-out validation sample(s). The prediction error rate is then estimated from the proportion of mistaken classifications among the total prediction events (Molinaro et al. 2005). When the principle rule of complete separation of the training data and test data is applied, each iteration has the potential to generate a different prediction model as it is developed from a different training set. Biases are inevitably introduced when combining the models for the purpose of developing the best model. Both split-sample and cross validation methods are generally discussed in the context of internal validation approaches – the validity of the developed predictive model can be best assessed when subjected to external validation using an entirely independent test set.

Class discovery aims to find expression patterns (i.e., groupings of samples or genes) that are previously unknown but once identified, can be assessed for correlation with known sample variables or for investigation of their biological or clinical implications. Class discovery methods are tightly related to data visualization methods. For example, the commonly used hierarchical clustering method groups genes and samples based on the overall similarity assessed by similarity measures such as correlation coefficient or Euclidean distance. Results of the clustering analysis are often displayed in a dendrogram format with a corresponding "heatmap" of expression values for each gene in each sample to facilitate visualization (Eisen et al. 1998). Clustering analysis is not a method for identifying differentially expressed genes but rather a tool used to reduce the high-dimension microarray data for assessing and visualizing the overall similarity among the genes and samples. Cluster analysis is generally referred in the context of unsupervised analysis in which the overall similarities (or distances) measured by the expression values of genes are independent of the sample labels (Eisen et al. 1998). However, the term "supervised clustering" is also used. Supervised clustering describes a method for visualizing the samples and gene relationship (Golub et al. 1999) based on selected genes from class comparison. As in any other steps in a microarray experiment, a variety of options can be considered in clustering analysis (Shannon et al. 2003). Since even random expression profiles can be clustered, it is essential to ascertain that the clustering results reflected the biologically meaningful clusters but not artifacts produced by the algorithm. The validity of the potential clusters can be tested using permutation-based approaches. The non-randomness of the cluster results can be validated if far more "strong classifiers" discriminating the clusters are found in the actual dataset than in the permutated dataset (Dougherty et al. 2002; McShane et al. 2002).

5.9 Clinical Translation of Microarray Studies

The ultimate validation of microarray derived study data is the demonstration of its utility in the clinical setting. One of the earliest examples is the identification of α-methylacyl-coA racemase (AMACR) as a robust prostate cancer marker. The potential of AMACR as a prostate cancer tissue marker was first proposed by Xu et al. (2000). AMACR, named P504S in the original publication, was identified by cDNA library subtraction in conjunction with high throughput microarray analysis of prostatic adenocarcinoma. However, the robustness of this cancer marker only came to light following large-scale array analysis that revealed its exceptionally high sensitivity and specificity as a prostate tumor marker (Luo et al. 2002; Rubin et al. 2002). Immunohistochemical detection of AMACR is a useful adjunctive method for detecting small foci of prostate cancer in prostate biopsies, which are performed one million times each year in the United States alone. As errors made in biopsy diagnosis either result in unnecessary surgery or delay of effective early treatment, the clinical utility of AMACR detection can not be underestimated. AMACR as a cancer marker has also been assessed in multiple organ sites (Went et al. 2006), and the underlying biology has spurred interest in novel strategies for cancer prevention, imaging, and treatment (Lloyd et al. 2008). A number of successes in breast cancer profiling studies (Sorlie et al. 2001; van't Veer et al. 2002; van de Vijver et al. 2002; Paik et al. 2004) have led to the commercialization of the use of expression profiles for breast cancer prognosis. Oncotype DX from Genomic Health (Sorlie et al. 2001; Cobleigh et al. 2005) is a multi-gene quantitative assay to assess the likelihood of breast cancer recurrence and benefit from chemotherapy. MammaPrint (van't Veer et al. 2002; van de Vijver et al. 2002) was developed in Europe as a 70-gene Agilent microarray based test for assessment of risk for breast cancer recurrence and have been approved by FDA in relatively early stage breast cancer patients. These examples and many others will continue to support the purported translational potential of study results derived from expression microarrays.

5.10 Future Outlook

Over the past few years, at least one aspect of standardization has been brought about by rapid advances in DNA microarray technology itself. Technical advances have been propelled by commercial interest and fueled by the continuously evolving annotation of the human genome as well as enhanced appreciation of the complex transcriptome. The number of transcripts in the mammalian genome is at least one order of magnitude greater than previously estimated because of the presence of alternative splicing, alternative transcriptional initiation and polyadenylation, transcription from both strands of DNA, and transcribed genomic fragments from regions of genome currently annotated as introns (Bertone et al. 2004; Carninci et al. 2005; Cheng et al. 2005). It is possible to design new arrays with probes targeting the entire transcriptome, or specialized arrays for detecting transcripts partitioned according to abundance to overcome the constraints posed by the relatively narrow dynamic range of the array technologies. A number of sequencing platforms simultaneously measure the sequence abundance and have the potential to reach wide-spread use for expression profiling once the cost is reduced (see Chap 6). The highly varied standard for selecting biological materials for expression analysis represents a major bottleneck in large-scale expression analysis (Tinker et al. 2006). The discovery of a clinically useful "predictor" of treatment outcome will depend on the availability of suitable patient samples with detailed clinical annotation including long term follow-up data. This bottleneck can be alleviated if archived formalin-fixed paraffin embedded (FFPE) pathological specimens can be routinely used in RNA-based analysis. Multiple parameters affecting RNA quality in FFPE specimens have been identified (von Ahlfen et al. 2007), raising the possibility of prospective archiving of FFPE specimens to minimize RNA degradation. With improved sampling and handling of clinical specimens, the next generation of expression analysis platforms will be more effective in addressing some of the most important questions in the field of cancer research.

References

Ahmed FE (2006) Microarray RNA transcriptional profiling: part I. Platforms, experimental design and standardization. Expert Rev Mol Diagn 6:535–550.

Attoor S, Dougherty ER, Chen Y, Bittner ML, Trent JM (2004) Which is better for cDNA-microarray-based classification: ratios or direct intensities. Bioinformatics 20:2513–2520.

Bammler T, Beyer RP, Bhattacharya S, Boorman GA, Boyles A, Bradford BU, Bumgarner RE, Bushel PR, Chaturvedi K, Choi D, Cunningham ML, Deng S, Dressman HK, Fannin RD,

Farin FM, Freedman JH, Fry RC, Harper A, Humble MC, Hurban P, Kavanagh TJ, Kaufmann WK, Kerr KF, Jing L, Lapidus JA, Lasarev MR, Li J, Li YJ, Lobenhofer EK, Lu X, Malek RL, Milton S, Nagalla SR, O'Malley JP, Palmer VS, Pattee P, Paules RS, Perou CM, Phillips K, Qin LX, Qiu Y, Quigley SD, Rodland M, Rusyn I, Samson LD, Schwartz DA, Shi Y, Shin JL, Sieber SO, Slifer S, Speer MC, Spencer PS, Sproles DI, Swenberg JA, Suk WA, Sullivan RC, Tian R, Tennant RW, Todd SA, Tucker CJ, Van Houten B, Weis BK, Xuan S, Zarbl H (2005) Standardizing global gene expression analysis between laboratories and across platforms. Nat Methods 2:351–356.

Barrett T, Edgar R (2006) Gene expression omnibus: microarray data storage, submission, retrieval, and analysis. Methods Enzymol 411:352–369.

Baugh LR, Hill AA, Brown EL, Hunter CP (2001) Quantitative analysis of mRNA amplification by in vitro transcription. Nucleic Acids Res 29:E29.

Bertone P, Stolc V, Royce TE, Rozowsky JS, Urban AE, Zhu X, Rinn JL, Tongprasit W, Samanta M, Weissman S, Gerstein M, Snyder M (2004) Global identification of human transcribed sequences with genome tiling arrays. Science 306:2242–2246.

Boelens MC, te Meerman GJ, Gibcus JH, Blokzijl T, Boezen HM, Timens W, Postma DS, Groen HJ, van den Berg A (2007) Microarray amplification bias: loss of 30% differentially expressed genes due to long probe – poly(A)-tail distances. BMC Genomics 8:277.

Bolstad BM, Irizarry RA, Astrand M, Speed TP (2003) A comparison of normalization methods for high density oligonucleotidde array data based on variance and bias. Bioinformatics 19:185–193.

Bovelstad HM, Nygard S, Storvold HL, Aldrin M, Borgan O, Frigessi A, Lingjaerde OC (2007) Predicting survival from microarray data–a comparative study. Bioinformatics 23:2080–2087.

Carbone M, Pass HI (2004) Multistep and multifactorial carcinogenesis: when does a contributing factor become a carcinogen? Semin Cancer Biol 14:399–405.

Carninci P, Kasukawa T, Katayama S, Gough J, Frith MC, Maeda N, Oyama R, Ravasi T, Lenhard B, Wells C, Kodzius R, Shimokawa K, Bajic VB, Brenner SE, Batalov S, Forrest AR, Zavolan M, Davis MJ, Wilming LG, Aidinis V, Allen JE, Ambesi-Impiombato A, Apweiler R, Aturaliya RN, Bailey TL, Bansal M, Baxter L, Beisel KW, Bersano T, Bono H, Chalk AM, Chiu KP, Choudhary V, Christoffels A, Clutterbuck DR, Crowe ML, Dalla E, Dalrymple BP, de Bono B, Della Gatta G, di Bernardo D, Down T, Engstrom P, Fagiolini M, Faulkner G, Fletcher CF, Fukushima T, Furuno M, Futaki S, Gariboldi M, Georgii-Hemming P, Gingeras TR, Gojobori T, Green RE, Gustincich S, Harbers M, Hayashi Y, Hensch TK, Hirokawa N, Hill D, Huminiecki L, Iacono M, Ikeo K, Iwama A, Ishikawa T, Jakt M, Kanapin A, Katoh M, Kawasawa Y, Kelso J, Kitamura H, Kitano H, Kollias G, Krishnan SP, Kruger A, Kummerfeld SK, Kurochkin IV, Lareau LF, Lazarevic D, Lipovich L, Liu J, Liuni S, McWilliam S, Madan Babu M, Madera M, Marchionni L, Matsuda H, Matsuzawa S, Miki H, Mignone F, Miyake S, Morris K, Mottagui-Tabar S, Mulder N, Nakano N, Nakauchi H, Ng P, Nilsson R, Nishiguchi S, Nishikawa S, Nori F, Ohara O, Okazaki Y, Orlando V, Pang KC, Pavan WJ, Pavesi G, Pesole G, Petrovsky N, Piazza S, Reed J, Reid JF, Ring BZ, Ringwald M, Rost B, Ruan Y, Salzberg SL, Sandelin A, Schneider C, Schonbach C, Sekiguchi K, Semple CA, Seno S, Sessa L, Sheng Y, Shibata Y, Shimada H, Shimada K, Silva D, Sinclair B, Sperling S, Stupka E, Sugiura K, Sultana R, Takenaka Y, Taki K, Tammoja K, Tan SL, Tang S, Taylor MS, Tegner J, Teichmann SA, Ueda HR, van Nimwegen E, Verardo R, Wei CL, Yagi K, Yamanishi H, Zabarovsky E, Zhu S, Zimmer A, Hide W, Bult C, Grimmond SM, Teasdale RD, Liu ET, Brusic V, Quackenbush J, Wahlestedt C, Mattick JS, Hume DA, Kai C, Sasaki D, Tomaru Y, Fukuda S, Kanamori-Katayama M, Suzuki M, Aoki J, Arakawa T, Iida J, Imamura K, Itoh M, Kato T, Kawaji H, Kawagashira N, Kawashima T, Kojima M, Kondo S, Konno H, Nakano K, Ninomiya N, Nishio T, Okada M, Plessy C, Shibata K, Shiraki T, Suzuki S, Tagami M, Waki K, Watahiki A, Okamura-Oho Y, Suzuki H, Kawai J, Hayashizaki Y (2005) The transcriptional landscape of the mammalian genome. Science 309:1559–1563.

Chen JJ, Wang SJ, Tsai CA, Lin CJ (2007) Selection of differentially expressed genes in microarray data analysis. Pharmacogenomics J 7:212–220.

Cheng J, Kapranov P, Drenkow J, Dike S, Brubaker S, Patel S, Long J, Stern D, Tammana H, Helt G, Sementchenko V, Piccolboni A, Bekiranov S, Bailey DK, Ganesh M, Ghosh S, Bell I, Gerhard DS, Gingeras TR (2005) Transcriptional maps of 10 human chromosomes at 5-nucleotide resolution. Science 308:1149–1154.

Cobleigh MA, Tabesh B, Bitterman P, Baker J, Cronin M, Liu ML, Borchik R, Mosquera JM, Walker MG, Shak S (2005) Tumor gene expression and prognosis in breast cancer patients with 10 or more positive lymph nodes. Clin Cancer Res 11:8623–8631.

Cox WG, Beaudet MP, Agnew JY, Ruth JL (2004) Possible sources of dye-related signal correlation bias in two-color DNA microarray assays. Anal Biochem 331:243–254.

Dobbin K, Simon R (2005) Sample size determination in microarray experiments for class comparison and prognostic classification. Biostatistics 6:27–38.

Dobbin KK, Beer DG, Meyerson M, Yeatman TJ, Gerald WL, Jacobson JW, Conley B, Buetow KH, Heiskanen M, Simon RM, Minna JD, Girard L, Misek DE, Taylor JM, Hanash S, Naoki K, Hayes DN, Ladd-Acosta C, Enkemann SA, Viale A, Giordano TJ (2005a) Interlaboratory comparability study of cancer gene expression analysis using oligonucleotide microarrays. Clin Cancer Res 11:565–572.

Dobbin KK, Kawasaki ES, Petersen DW, Simon RM (2005b) Characterizing dye bias in microarray experiments. Bioinformatics 21:2430–2437.

Dougherty ER, Barrera J, Brun M, Kim S, Cesar RM, Chen Y, Bittner M, Trent JM (2002) Inference from clustering with application to gene-expression microarrays. J Comput Biol 9:105–126.

Edgar R, Domrachev M, Lash AE (2002) Gene expression omnibus: NCBI gene expression and hybridization array data repository. Nucleic Acids Res 30:207–210.

Eisen MB, Spellman PT, Brown PO, Botstein D (1998) Cluster analysis and display of genome-wide expression patterns. Proc Natl Acad Sci U S A 95:14863–14868.

Elvidge G (2006) Microarray expression technology: from start to finish. Pharmacogenomics 7:123–134.

Fan W, Pritchard JI, Olson JM, Khalid N, Zhao LP (2005) A class of models for analyzing GeneChip gene expression analysis array data. BMC Genomics 6:16.

Golub TR, Slonim DK, Tamayo P, Huard C, Gaasenbeek M, Mesirov JP, Coller H, Loh ML, Downing JR, Caligiuri MA, Bloomfield CD, Lander ES (1999) Molecular classification of cancer: class discovery and class prediction by gene expression monitoring. Science 286:531–537.

Hanahan D, Weinberg RA (2000) The hallmarks of cancer. Cell 100:57–70.

Hardiman G (2004) Microarray platforms – comparisons and contrasts. Pharmacogenomics 5:487–502.

Harr B, Schlotterer C (2006) Comparison of algorithms for the analysis of Affymetrix microarray data as evaluated by co-expression of genes in known operons. Nucleic Acids Res 34:e8.

Hughes TR, Mao M, Jones AR, Burchard J, Marton MJ, Shannon KW, Lefkowitz SM, Ziman M, Schelter JM, Meyer MR, Kobayashi S, Davis C, Dai H, He YD, Stephaniants SB, Cavet G, Walker WL, West A, Coffey E, Shoemaker DD, Stoughton R, Blanchard AP, Friend SH, Linsley PS (2001) Expression profiling using microarrays fabricated by an ink-jet oligonucleotide synthesizer. Nat Biotechnol 19:342–347.

Irizarry RA, Warren D, Spencer F, Kim IF, Biswal S, Frank BC, Gabrielson E, Garcia JG, Geoghegan J, Germino G, Griffin C, Hilmer SC, Hoffman E, Jedlicka AE, Kawasaki E, Martinez-Murillo F, Morsberger L, Lee H, Petersen D, Quackenbush J, Scott A, Wilson M, Yang Y, Ye SQ, Yu W (2005) Multiple-laboratory comparison of microarray platforms. Nat Methods 2:345–350.

Jeffery IB, Higgins DG, Culhane AC (2006) Comparison and evaluation of methods for generating differentially expressed gene lists from microarray data. BMC Bioinformatics 7:359.

Jiang Z, Gentleman R (2007) Extensions to gene set enrichment. Bioinformatics 23:306–313.

Jin W, Riley RM, Wolfinger RD, White KP, Passador-Gurgel G, Gibson G (2001) The contributions of sex, genotype and age to transcriptional variance in *Drosophila melanogaster*. Nat Genet 29:389–395.

Kong SW, Pu WT, Park PJ (2006) A multivariate approach for integrating genome-wide expression data and biological knowledge. Bioinformatics 22:2373–2380.

Kuo WP, Liu F, Trimarchi J, Punzo C, Lombardi M, Sarang J, Whipple ME, Maysuria M, Serikawa K, Lee SY, McCrann D, Kang J, Shearstone JR, Burke J, Park DJ, Wang X, Rector TL, Ricciardi-Castagnoli P, Perrin S, Choi S, Bumgarner R, Kim JH, Short GF 3rd, Freeman MW, Seed B, Jensen R, Church GM, Hovig E, Cepko CL, Park P, Ohno-Machado L, Jenssen TK (2006) A sequence-oriented comparison of gene expression measurements across different hybridization-based technologies. Nat Biotechnol 24:832–840.

Lee JC, Stiles D, Lu J, Cam MC (2007) A detailed transcript-level probe annotation reveals alternative splicing based microarray platform differences. BMC Genomics 8:284.

Lloyd MD, Darley DJ, Wierzbicki AS, Threadgill MD (2008) Alpha-methylacyl-CoA racemase – an 'obscure' metabolic enzyme takes centre stage. FEBS J 275:1089–1102.

Luo J, Duggan DJ, Chen Y, Sauvageot J, Ewing CM, Bittner ML, Trent JM, Isaacs WB (2001) Human prostate cancer and benign prostatic hyperplasia: molecular dissection by gene expression profiling. Cancer Res 61:4683–4688.

Luo J, Zha S, Gage WR, Dunn TA, Hicks JL, Bennett CJ, Ewing CM, Platz EA, Ferdinandusse S, Wanders RJ, Trent JM, Isaacs WB, De Marzo AM (2002) Alpha-methylacyl-CoA racemase: a new molecular marker for prostate cancer. Cancer Res 62:2220–2226.

Luo J, Isaacs WB, Trent JM, Duggan DJ (2003) Looking beyond morphology: cancer gene expression profiling using DNA microarrays. Cancer Invest 21:937–949.

McShane LM, Radmacher MD, Freidlin B, Yu R, Li MC, Simon R (2002) Methods for assessing reproducibility of clustering patterns observed in analyses of microarray data. Bioinformatics 18:1462–1469.

Molinaro AM, Simon R, Pfeiffer RM (2005) Prediction error estimation: a comparison of resampling methods. Bioinformatics 21:3301–3307.

Nelson WG, De Marzo AM, Isaacs WB (2003) Prostate cancer. N Engl J Med 349:366–381.

Paik S, Shak S, Tang G, Kim C, Baker J, Cronin M, Baehner FL, Walker MG, Watson D, Park T, Hiller W, Fisher ER, Wickerham DL, Bryant J, Wolmark N (2004) A multigene assay to predict recurrence of tamoxifen-treated, node-negative breast cancer. N Engl J Med 351:2817–2826.

Patterson TA, Lobenhofer EK, Fulmer-Smentek SB, Collins PJ, Chu TM, Bao W, Fang H, Kawasaki ES, Hager J, Tikhonova IR, Walker SJ, Zhang L, Hurban P, de Longueville F, Fuscoe JC, Tong W, Shi L, Wolfinger RD (2006) Performance comparison of one-color and two-color platforms within the MicroArray Quality Control (MAQC) project. Nat Biotechnol 24:1140–1150.

Pease AC, Solas D, Sullivan EJ, Cronin MT, Holmes CP, Fodor SP (1994) Light-generated oligonucleotide arrays for rapid DNA sequence analysis. Proc Natl Acad Sci U S A 91:5022–5026.

Quackenbush J (2002) Microarray data normalization and transformation. Nat Genet 32(Suppl):496–501.

Radmacher MD, McShane LM, Simon R (2002) A paradigm for class prediction using gene expression profiles. J Comput Biol 9:505–511.

Ransohoff DF (2004) Rules of evidence for cancer molecular-marker discovery and validation. Nat Rev Cancer 4:309–314.

Reiner A, Yekutieli D, Benjamini Y (2003) Identifying differentially expressed genes using false discovery rate controlling procedures. Bioinformatics 19:368–375.

Rosenzweig BA, Pine PS, Domon OE, Morris SM, Chen JJ, Sistare FD (2004) Dye bias correction in dual-labeled cDNA microarray gene expression measurements. Environ Health Perspect 112:480–487.

Rubin MA, Zhou M, Dhanasekaran SM, Varambally S, Barrette TR, Sanda MG, Pienta KJ, Ghosh D, Chinnaiyan AM (2002) alpha-Methylacyl coenzyme A racemase as a tissue biomarker for prostate cancer. JAMA 287:1662–1670.

Schena M, Shalon D, Davis RW, Brown PO (1995) Quantitative monitoring of gene expression patterns with a complementary DNA microarray. Science 270:467–470.

Shannon W, Culverhouse R, Duncan J (2003) Analyzing microarray data using cluster analysis. Pharmacogenomics 4:41–52.

Shi L, Reid LH, Jones WD, Shippy R, Warrington JA, Baker SC, Collins PJ, de Longueville F, Kawasaki ES, Lee KY, Luo Y, Sun YA, Willey JC, Setterquist RA, Fischer GM, Tong W, Dragan YP, Dix DJ, Frueh FW, Goodsaid FM, Herman D, Jensen RV, Johnson CD, Lobenhofer EK, Puri RK, Schrf U, Thierry-Mieg J, Wang C, Wilson M, Wolber PK, Zhang L, Amur S, Bao W, Barbacioru CC, Lucas AB, Bertholet V, Boysen C, Bromley B, Brown D, Brunner A, Canales R, Cao XM, Cebula TA, Chen JJ, Cheng J, Chu TM, Chudin E, Corson J, Corton JC, Croner LJ, Davies C, Davison TS, Delenstarr G, Deng X, Dorris D, Eklund AC, Fan XH, Fang H, Fulmer-Smentek S, Fuscoe JC, Gallagher K, Ge W, Guo L, Guo X, Hager J, Haje PK, Han J, Han T, Harbottle HC, Harris SC, Hatchwell E, Hauser CA, Hester S, Hong H, Hurban P, Jackson SA, Ji H, Knight CR, Kuo WP, LeClerc JE, Levy S, Li QZ, Liu C, Liu Y, Lombardi MJ, Ma Y, Magnuson SR, Maqsodi B, McDaniel T, Mei N, Myklebost O, Ning B, Novoradovskaya N, Orr MS, Osborn TW, Papallo A, Patterson TA, Perkins RG, Peters EH, Peterson R, Philips KL, Pine PS, Pusztai L, Qian F, Ren H, Rosen M, Rosenzweig BA, Samaha RR, Schena M, Schroth GP, Shchegrova S, Smith DD, Staedtler F, Su Z, Sun H, Szallasi Z, Tezak Z, Thierry-Mieg D, Thompson KL, Tikhonova I, Turpaz Y, Vallanat B, Van C, Walker SJ, Wang SJ, Wang Y, Wolfinger R, Wong A, Wu J, Xiao C, Xie Q, Xu J, Yang W, Zhang L, Zhong S, Zong Y, Slikker W Jr (2006) The MicroArray Quality Control (MAQC) project shows inter- and intraplatform reproducibility of gene expression measurements. Nat Biotechnol 24:1151–1161.

Simon R, Radmacher MD, Dobbin K, McShane LM (2003) Pitfalls in the use of DNA microarray data for diagnostic and prognostic classification. J Natl Cancer Inst 95:14–18.

Smyth GK, Yang YH, Speed T (2003) Statistical issues in cDNA microarray data analysis. Methods Mol Biol 224:111–136.

Sorlie T, Perou CM, Tibshirani R, Aas T, Geisler S, Johnsen H, Hastie T, Eisen MB, van de Rijn M, Jeffrey SS, Thorsen T, Quist H, Matese JC, Brown PO, Botstein D, Eystein Lonning P, Borresen-Dale AL (2001) Gene expression patterns of breast carcinomas distinguish tumor subclasses with clinical implications. Proc Natl Acad Sci U S A 98:10869–10874.

Southern EM, Maskos U, Elder JK (1992) Analyzing and comparing nucleic acid sequences by hybridization to arrays of oligonucleotides: evaluation using experimental models. Genomics 13:1008–1017.

Storey JD, Xiao W, Leek JT, Tompkins RG, Davis RW (2005) Significance analysis of time course microarray experiments. Proc Natl Acad Sci U S A 102:12837–12842.

Subramanian A, Tamayo P, Mootha VK, Mukherjee S, Ebert BL, Gillette MA, Paulovich A, Pomeroy SL, Golub TR, Lander ES, Mesirov JP (2005) Gene set enrichment analysis: a knowledge-based approach for interpreting genome-wide expression profiles. Proc Natl Acad Sci U S A 102:15545–15550.

Taylor J, Tibshirani R, Efron B (2005) The 'miss rate' for the analysis of gene expression data. Biostatistics 6:111–117.

Thompson KL, Pine PS, Rosenzweig BA, Turpaz Y, Retief J (2007) Characterization of the effect of sample quality on high density oligonucleotide microarray data using progressively degraded rat liver RNA. BMC Biotechnol 7:57.

Tian L, Greenberg SA, Kong SW, Altschuler J, Kohane IS, Park PJ (2005) Discovering statistically significant pathways in expression profiling studies. Proc Natl Acad Sci U S A 102:13544–13549.

Tinker AV, Boussioutas A, Bowtell DD (2006) The challenges of gene expression microarrays for the study of human cancer. Cancer Cell 9:333–339.

Tusher VG, Tibshirani R, Chu G (2001) Significance analysis of microarrays applied to the ionizing radiation response. Proc Natl Acad Sci U S A 98:5116–5121.

van de Vijver MJ, He YD, van't Veer LJ, Dai H, Hart AA, Voskuil DW, Schreiber GJ, Peterse JL, Roberts C, Marton MJ, Parrish M, Atsma D, Witteveen A, Glas A, Delahaye L, van der Velde T, Bartelink H, Rodenhuis S, Rutgers ET, Friend SH, Bernards R (2002) A gene-expression signature as a predictor of survival in breast cancer. N Engl J Med 347:1999–2009.

Van Gelder RN, von Zastrow ME, Yool A, Dement WC, Barchas JD, Eberwine JH (1990) Amplified RNA synthesized from limited quantities of heterogeneous cDNA. Proc Natl Acad Sci U S A 87:1663–1667.

van't Veer LJ, Dai H, van de Vijver MJ, He YD, Hart AA, Mao M, Peterse HL, van der Kooy K, Marton MJ, Witteveen AT, Schreiber GJ, Kerkhoven RM, Roberts C, Linsley PS, Bernards R, Friend SH (2002) Gene expression profiling predicts clinical outcome of breast cancer. Nature 415:530–536.

von Ahlfen S, Missel A, Bendrat K, Schlumpberger M (2007) Determinants of RNA quality from FFPE samples. PLoS ONE 2:e1261.

Walt DR (2000) Techview: molecular biology. Bead-based fiber-optic arrays. Science 287: 451–452.

Wang E, Miller LD, Ohnmacht GA, Liu ET, Marincola FM (2000) High-fidelity mRNA amplification for gene profiling. Nat Biotechnol 18:457–459.

Weis S, Llenos IC, Dulay JR, Elashoff M, Martinez-Murillo F, Miller CL (2007) Quality control for microarray analysis of human brain samples: the impact of postmortem factors, RNA characteristics, and histopathology. J Neurosci Methods 165:198–209.

Went PT, Sauter G, Oberholzer M, Bubendorf L (2006) Abundant expression of AMACR in many distinct tumour types. Pathology 38:426–432.

Wolber PK, Collins PJ, Lucas AB, De Witte A, Shannon KW (2006) The Agilent in situ-synthesized microarray platform. Methods Enzymol 410:28–57.

Wu Z, Irizarry RA (2004) Preprocessing of oligonucleotide array data. Nat Biotechnol 22:656–658; author reply 658.

Xu J, Stolk JA, Zhang X, Silva SJ, Houghton RL, Matsumura M, Vedvick TS, Leslie KB, Badaro R, Reed SG (2000) Identification of differentially expressed genes in human prostate cancer using subtraction and microarray. Cancer Res 60:1677–1682.

Yang MC, Yang JJ, McIndoe RA, She JX (2003) Microarray experimental design: power and sample size considerations. Physiol Genomics 16:24–28.

Chapter 6
Signal Sequencing for Gene Expression Profiling

Biaoyang Lin, Jeremy Wechsler, and Leroy Hood

Abstract Over the past decade, advances in DNA sequencing technologies have made sequencing entire genomes a reality. The ever-expanding size and detail of the genomic data has created a solid framework for the rapid development of sensitive, high throughput gene expression profiling techniques. In this chapter, we discuss, in detail, the ways in which SAGE and MPSS signal sequencing methods have been used to conduct thorough comparative gene expression profiles, the advantages these methods have over traditional expression profiling techniques (i.e. microarrays), and their potential to significantly contribute to understanding the perturbed signaling networks of cancer. Because there are many factors that greatly influence the quality of the data produced by sequencing based expression profiling, the specifics of approaches used in data analysis and the factors to consider when mapping signal sequence data to the transcriptome or genome are presented to, hopefully, help researchers in their current and future gene expression profiling research. We use gene expression data from prostate and ovarian cancer to illustrate the power these technologies hold for generating "deep" and sensitive (i.e. a wide dynamic range) expression profiles, and, finally, we discuss the development of the next generation of sequencing technologies and their application to deciphering the cancer transcriptome. High throughput technologies coupled with a broad, systems-based approach to understanding disease will substantially aid in the development of clinical tools for disease diagnosis and prognosis and will undoubtedly contribute to the design of novel and efficacious therapeutics.

B. Lin (✉)
Department of Urology, University of Washington, Seattle, WA, USA
Zhejiang-California International Nanosystems Institute, Hangzhou China
e-mail: bylin@u.washington.edu

S. Yegnasubramanian and W.B. Isaacs (eds.), *Modern Molecular Biology*, Applied Bioinformatics and Biostatistics in Cancer Research,
DOI 10.1007/978-0-387-69745-1_6,

6.1 Introduction

Signal sequencing technologies offer significant advantages over DNA microarray technologies in gene expression profiling, although they were developed in parallel. The variable formats of microarrays, such as cDNA microarrays and oligo microarrays, have played important roles in gene expression profiling. Despite their utility, however, microarrays have several limitations: (1) coverage of the transcriptome depends on the availability of cDNA sequences or detailed genome annotations to design appropriate oligonucleotide probes; (2) the sensitivity is limited because it is an analog based technology subject to background noise. In our experience, only 25–50% of an entire transcriptome array gives reliable signals (e.g. signal intensities twofold greater than background). Applying a filter to eliminate signals that are too close to background levels can help, but they have not always been used and, as a result, some of the earliest microarray studies may have drawn questionable conclusions; (3) data obtained from different microarray formats (i.e. cDNA, printed oligo array, in situ synthesis oligo array, etc.) often does not overlap. The lack of reproducibility may be due, in part, to slight reagent differences and the intrinsic properties of the probes (e.g. oligo length and hybridization kinetics); (4) microarrays are limited by their dynamic range. Highly expressed genes are easily saturated, while genes only moderately expressed can be obscured by background signals; (5) even when using short oligos, microarrays suffer from cross-hybridization and, unfortunately, the true and specific hybridization intensity of a probe is confounded by cross-hybridization signals.

Sequencing-based technology has transformed expression profiling, especially with the advent of next generation sequencing technologies. The major advantage of signal sequencing based expression profiling, when compared to microarray, is that it does not rely on the existence of a physical piece of DNA (e.g. amplified inserts from cDNA libraries) or oligos designed and synthesized from known gene/wtranscript annotations. Although rapid advances in DNA sequencing technologies have made sequencing entire genomes a reality (>1,700 genomes have already been sequenced and deposited at NCBI), genome annotation (i.e. identification of exons, entire mRNA transcripts and more obscure transcript types, such as miRNA and ncRNA) has, understandably, been unable to keep pace. Microarrays, therefore, are limited by these incomplete genome annotations and currently are only possible for partially annotated species, such as human, mouse, yeast, and rice. Because sequence-based technology does not rely on pre-existing genome or transcript annotation, it is both an expression profiling tool for known transcripts and a discovery tool to identify novel transcripts. This technology holds great promise and may eventually contribute to genome annotations.

Another advantage of signal sequencing based expression profiling is that it is a counting based technology, making it truly digital in nature, compared to the analog nature of the signal output (e.g. fluorescence) in microarrays. The signal can be expressed in normalized units of transcript per million (tpm) and it makes comparisons between different samples relatively simple. This circumvents the problems of data normalization, reproducibility, and data comparability between labs and different platforms that have plagued microarray studies.

Two factors have the greatest influence on the success of signal sequencing based expression profiling. First, the sequence length for each tag greatly affects the specificity and sensitivity of the procedure to mapping the sequenced tags to the transcripts they represent. Second, the quality and completeness of the sample sequencing (i.e. the "depth" of sequencing a sample) will obviously have a large impact on the data output With the arrival of high throughput next generation sequencing technologies, we will now be able to sequence millions of tags cheaply, thereby achieving increased dynamic range and detection sensitivity. Using our estimate that the total number of mRNA molecules (the number of different mRNA species times number of molecules for each mRNA species) in a typical sample is in the range of 300–500 K (Lin et al. 2005), a sequence depth of a few million tags will allow us to detect mRNA expressed at less than one copy per cell (i.e. not all cells express the mRNA). This ability "to see all things" is aligned with a systems biology approach to understanding biological pathways. In fact, the first prerequisite of systems biology is to enumerate all components in a system.

In this chapter, we will focus on sequencing based technologies for expression profiling. The methods discussed include SAGE, MPSS, and some of the newer, next generation, sequencing technologies such as SBS (sequencing by synthesis) and their application to cancer research.

6.2 Technologies for Generating Signal Sequences for Expression Profiling

6.2.1 Serial Analysis of Gene Expression

The earliest attempts to use counting based technology to detect differences in gene expression probably came from the so-called "digital Northern" or "electronic Northern" analysis. Electronic Northern is based on comparing the number of ESTs (Expressed Sequence Tags) for a gene identified in different cDNA libraries (e.g. from the EST databases) to be compared. However, because the sequence depth (number of ESTs sequenced in a library) of most cDNA libraries is very limited (often thousands of ESTs), the information content is limited. Another issue is that many cDNA libraries are "normalized libraries" or "substracted libraries" and therefore could not be used for electronic Northern blot analysis.

Serial analysis of gene expression (SAGE) is the first true counting based (counting the tags sequenced) gene expression profiling technology developed by Velculescu et al. in 1995 (Velculescu et al. 1995). They cleverly employed a type II restriction enzyme that cleaves a defined distance downstream from its recognition sequence to generate a 14bp SAGE tag, which is sufficiently long enough to reliably identify transcripts. The other component of the procedure that made SAGE

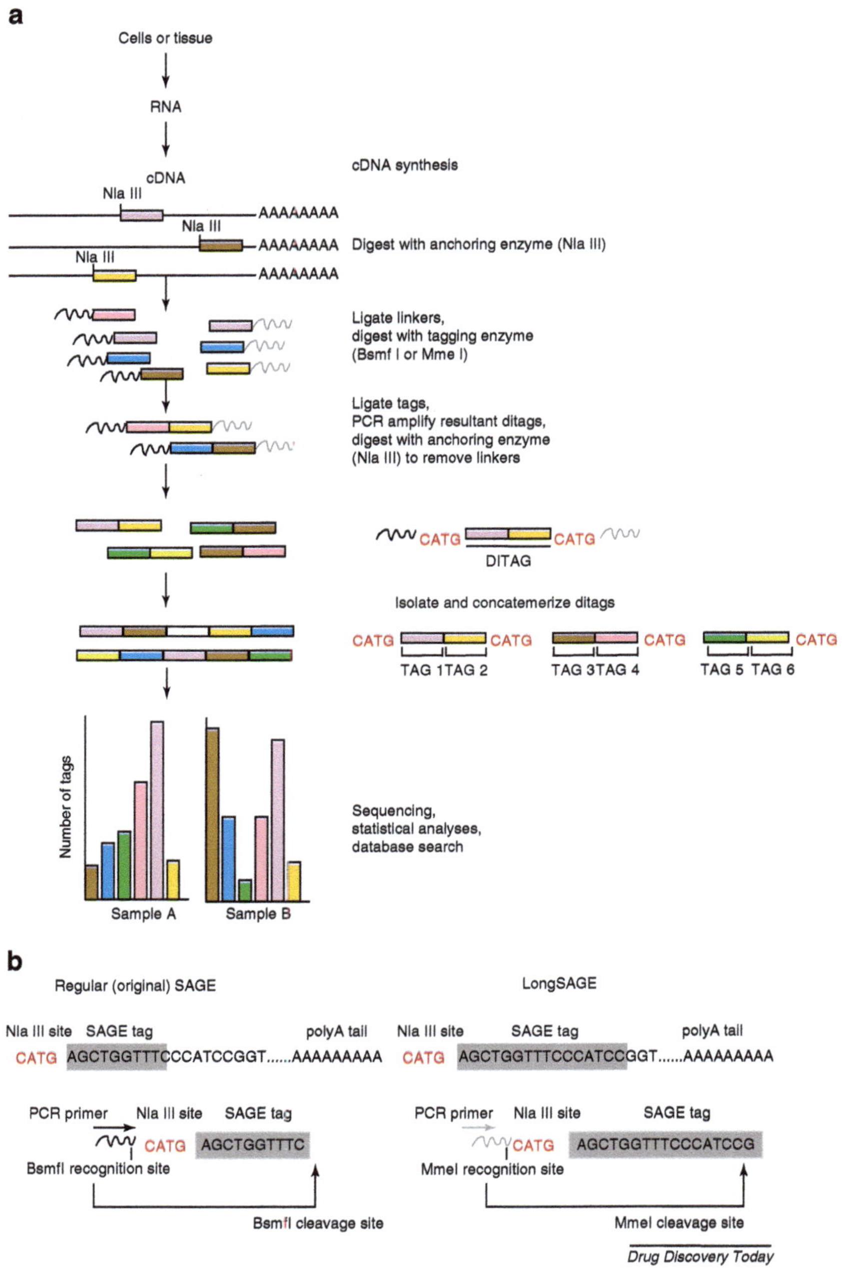

Fig. 6.1 Outline of the SAGE method. (**a**) Schematic outline of SAGE library generation and analysis. Double stranded cDNA is synthesized from mRNA isolated from cells or tissues and immobilized to oligo(dT)-magnetic beads, and then digested using the anchoring enzyme (commonly Nla III). Following Nla III digestion, linkers that contain a recognition site for the tagging enzyme (Bsmf I for regular SAGE or Mme I for LongSAGE) are ligated to the 3′ cDNA ends.

possible is the concatemerization of the SAGE tags into longer pieces. This enabled efficient sequencing using standard automatic sequencing technology available at the time, in the mid-1990s.

An outline of the SAGE procedure is included below (Fig 6.1) and standard SAGE protocols are available online (e.g. SAGENET: www.sagenet.org, MD Anderson SAGE site: sciencepark.mdanderson.org/ggeg/default.html; www.protocol-online.org). Commercial kits and protocols for generating SAGE libraries are also available (e.g. I-SAGE kit and I-SAGE long kit from Invitrogen Inc.).

Because the original SAGE procedure (Velculescu et al. 1995) used the BsmF I type IIs enzyme as the "tagging enzyme" that cuts approximately 12 bp downstream of its recognition site (BsmF I was predicted to digest 14 bp downstream of its recognition site in its native 65°C digestion temperature. However, it was used at 37°C in the SAGE protocol, which was found to cleave about 12 bp downstream of its recognition site), it can only reliably generate a SAGE tag of 9 bp (in combination with NlaIII "anchoring enzyme"). Interestingly, although Velculescu et al. only took 9 bp sequences as SAGE tags, many others took 10 bp sequences as the SAGE tags, even though the same protocol was used (see the NCBI Geo database). A tag length of 14 bp is routinely reported for SAGE tags because the 4 bp recognition site of NlaIII (CATG) is included.

When novel type IIs and type III restriction enzymes that cut farther from their recognition sites became available, it was possible to generate longer SAGE tags. A longer tag was preferable because it allowed one to more reliably map the tag to a specific transcript. Indeed, Saha et al. created a method to generate 21 bp SAGE tags using the Mme I type II restriction enzyme, appropriately named the LongSAGE method (Saha et al. 2002). Furthermore, Matsumura et al. developed a method for generating 27 bp SAGE tags, the so-called SuperSAGE method, which employed EcoP15I, a type III restriction enzyme (Matsumura et al. 2003).

Fig. 6.1 (continued) Linker-tag fragments are then released from the cDNA following digestion with the tagging enzyme. Resulting free linker-tag fragments are ligated together into "ditags", PCR amplified, concatemerized, subcloned into a vector and finally sequenced as one long fragment of DNA. Each 14 bp (regular SAGE) or 21 bp (LongSAGE) tag should uniquely identify a specific gene transcript and the abundance of tags sequenced in a given library reflects the absolute transcript level within the sample analyzed. SAGE tags are indicated with differently colored bars, whereas wavy black and gray lines denote linkers. (**b**) Comparison of regular (original) and LongSAGE. Only the step that is different between the two procedures is highlighted. In the case of regular SAGE, the tagging enzyme is Bsmf I that will lead to the generation of 14 bp tags (CATG+10 nucleotides), whereas in LongSAGE, the use of Mme I as tagging enzyme will result in 21 bp tags (CATG+17 nucleotides). Location of the Nla III sites (CATG), tagging enzyme recognition sites (in the linkers) and cleavage sites, PCR primers within the linkers (*wavy black* and *gray lines*) and sequence of potential tags (*highlighted in gray*) are shown) (reproduced from Porter et al. (2006) with permission from Elsevier)

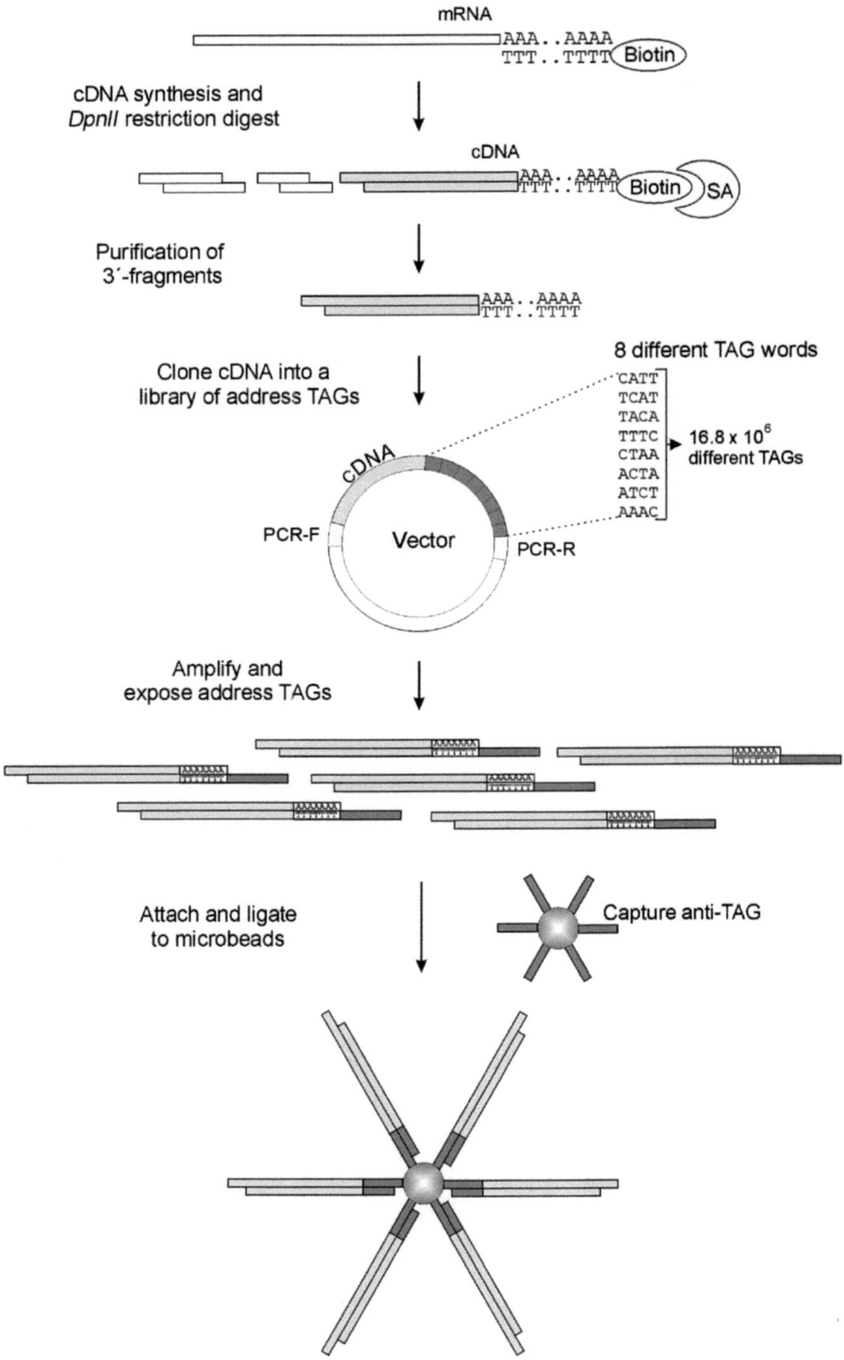

Fig. 6.2 Preparation of cDNA loaded microbeads. ES cell poly(A+) mRNA is converted into double-stranded cDNA, which is digested with Dpn II, followed by capture of the 3′-most DpnII

6.2.2 Massively Parallel Signature Sequencing

The development of efficient and cheap next generation sequencing technology were necessary to address the issue that only a fraction of the tags were actually being sequenced from the total transcript population with the SAGE technology (i.e. lack of sequence depth or coverage).

The MPSS technology was invented by Sydney Brenner (Brenner et al. 2000a, b) and commercialized by Lynx therapeutics Inc. (later Solexa Inc). The generation of tags for sequencing is very similar to that for SAGE (see the MPSS method description below). There are two main advantages of MPSS over SAGE. First, it employs a delicate sequence by ligation scheme that generates tags up to 20 bp, making MPSS more specific and reliable than the traditional SAGE method. Second, and inarguably the major advantage of MPSS over SAGE, is the depth of sequence coverage. Two million tags are routinely sequenced compared to the 100 K typical for SAGE analysis. This permits the more reliable detection of low abundance transcripts and gives the data much higher statistical power for the comparative expression profiling analysis.

The MPSS procedure is illustrated in Figs. 6.2 and 6.3. In brief, the procedure is generally carried out as follows: after preparing an mRNA sample, cDNA is generated using labeled poly(A) primers (e.g. biotin). The pool of cDNA is then digested with DpnII, creating cDNA fragments with GATC overhangs from the cleavage sites closest to their 3′ end. All these biotin-labeled cDNA fragments are then purified with streptavidin beads. Next, adaptor molecules containing a specific type II restriction enzyme (e.g. Mme I) recognition sequence are ligated to the 5′end (GATC overhang end) of the bound cDNA fragments (Fig. 6.2). Subsequent cleavage with a type II enzyme, such as Mme I, which cuts 20–21 bps downstream from its recognition sequence, releases the original adaptor. This released fragment, however, now contains a short portion of the original cDNA. A second adaptor molecule is now ligated to its 3′end. The resulting cDNA fragments are then cloned into a vector containing "Combi tags" (Daixing Zhou, personal communication) generated by combinatorial synthesis of 32 nucleotides. There are approximately 16.8 million different Combi tags. Lynx therapeutics Inc. (now part of Illumina Inc.)

Fig. 6.2 (continued) fragments. These are converted to 20 base inserts flanked by the adapters to allow cloning into a plasmid containing a 32-basepair oligonucleotide tag. Each cDNA clone is associated with one of 16.7 million different 32 base tag sequences. The cDNA inserts, along with their associated tags, are amplified by polymerase chain reaction, and the resulting amplicons are treated with an exonuclease to render the tag portions single stranded. The tagged cDNAs are hybridized to 32-base complementary tags that are covalently linked to the microbeads. For every tag on a cDNA molecule, there is one bead with the complementary anti-tag. After loading, the tagged cDNAs are then ligated enzymatically to yield a microbead with approximately 100,000 identical molecules covalently attached to the surface (reproduced from Reinartz et al. (2002) by permission of Oxford University Press)

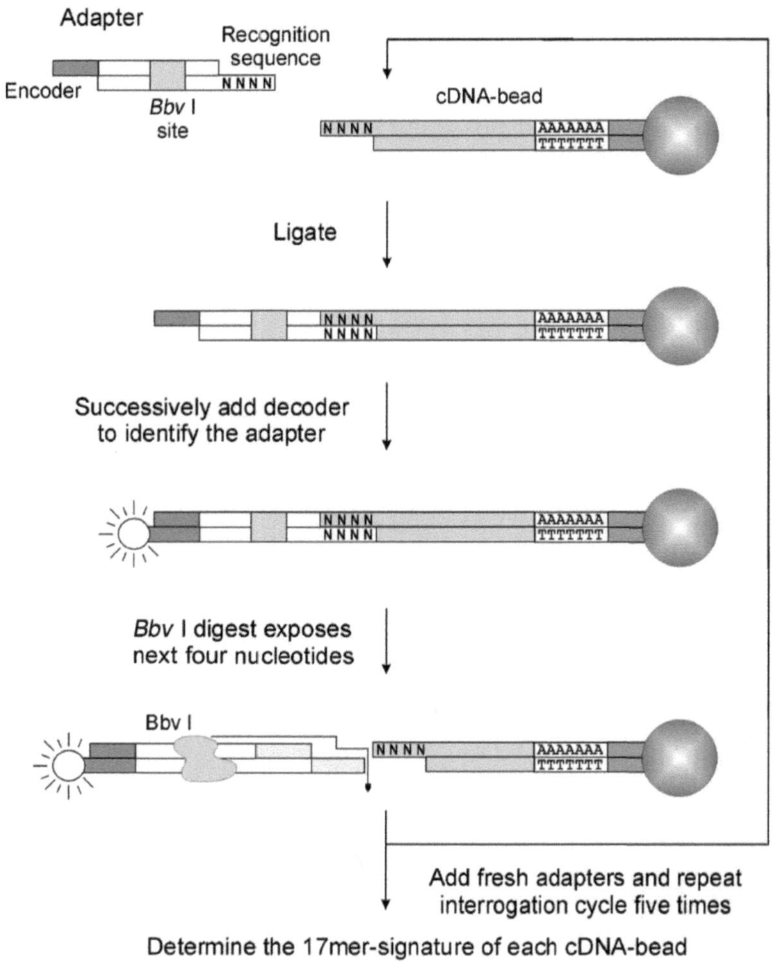

Fig. 6.3 Sequencing by adapter ligation cycling. Encoded adapters are ligated to the four base single-stranded overhangs at the end of the cDNAs attached to microbeads. Sixteen different fluorescently labeled decoder probes are sequentially hybridized to the ends of the encoded adapters to identify the sequence of the exposed four bases. The encoded adapter from the first round is then removed by digestion with BbvI, which exposes the next four nucleotides as a four-base single-stranded overhang. Repetition of the process yields up to 20 bases of sequence (reproduced from Reinartz et al. (2002) by permission of Oxford University Press)

created "megaclones" (Fig. 6.3), in which 5-um size microbeads are pre-conjugated with sequences complementary to the Combi tags. After hybridization with the tagged cDNAs, each microbead retains 100 K copies of cDNAs. The microbeads are then loaded into a microfluidic flow cell for sequencing, in which they form a monolayer along the microfluidic channel. It is a ligation mediated sequencing process and is described in detail by Brenner et al. in the original publication (Brenner et al. 2000a).

The protocol we described above is named "signature cloning" (Daixing Zhou, personal communication) as opposed to the original protocol that was referred to as "classic cloning" (Brenner et al. 2000b). The difference is that only 20–21 bps after the last Dpn II site are cloned in the signature cloning protocol, while the entire piece of cDNA from the last Dpn II site to the poly A tail is cloned in the classic cloning. Classic cloning protocol is more prone to bias during the construction of tag libraries due to different amplification and cloning efficiency among cDNAs with different lengths from the last Dpn II site to their poly A tails.

6.3 Data Analysis for Signal Sequencing Based Expression Profiling

6.3.1 Mapping Tag to Gene

The first step to analyze the signal sequencing data, generated by any of the methods discussed above, is to map the tag sequences to the transcriptome. There are two main approaches to map tags. The first, and most obvious way, is to do a database search directly using a BLAST search tool or another fast sequence alignment algorithm, such as the ELAND program by Illumina/Solexa Inc. The sheer quantity of unique tags to analyze, however, can make direct database searches time consuming, and although they are, when comparing samples, capable of producing lists of differentially expressed genes, they do not efficiently extract reliable biological information. By pre-constructing an annotated virtual tag database, based on all known transcripts (ESTs and genes) and genomic sequences, the positional information of the tag (e.g. the last NlaIII site for SAGE tags or last DpnII site before the polyA tail for MPSS tags) and the orientation of a tag against a transcript (e.g. whether or not a matched transcript has a poly A tail) can be taken into account. For analyzing SAGE data, pre-constructed virtual database and algorithms were available: SAGEmap (Lal et al. 1999) (http://www.ncbi.nlm.nih.gov/projects/SAGE/), SAGE Genie (Boon et al. 2002) (http://cgap.nci.nih.gov/SAGE), and TAGmapper (Bala et al. 2005) (http://tagmapper.ibioinformatics.org).

Lynx has also created a virtual MPSS tag database. These virtual tags can then be categorized into different classes and types based on their characteristics, such as the position and orientation of a MPSS tag against a transcript and whether or not the matched transcript has a poly A tail and/or a polyadenylation signal sequence. We have applied Lynx's virtual MPSS database in mapping our prostate cancer MPSS data (Lin et al. 2005) and Table 6.1 lists major categories used to classify MPSS tags (reproduced from supplementary table S1 from our previous publication (Lin et al. 2005). Class1–3 MPSS tags represent the most reliable identifiers of their corresponding transcripts (Table 6.1). To map the tags from custom MPSS datasets, they were directly compared to the virtual MPSS tags. The transcripts identified by the virtual tags were then retrieved to represent our tags.

Unigene has done a tremendous job clustering transcripts and ESTs and has built an excellent database with 122,083 entries for the human transcriptome

Table 6.1 Classification of MPSS signatures

Signature class	Class definition	Poly A signal	Poly A tail	Strand
1	1. The signature is seen in the forward strand of the cDNA 2. The cDNA has a polyA signal and a polyA tail 3. The signature is the first one 5′ to the polyA signal and the polyA tail	Yes	Yes	Forward
2	1. The signature is seen in the forward strand of the cDNA 2. The cDNA has a polyA signal but no polyA tails 3. The signature is the first one 5′ to the polyA signal	Yes	No	Forward
3	1. The signature is seen in the forward strand of the cDNA 2. The cDNA has a polyA tail but no polyA signals 3. The signature is the first one 5′ to the polyA tail	No	Yes	Forward
4	1. The signature is seen in the forward strand of the cDNA 2. The cDNA has no polyA signals and no polyA tail 3. The signature is the most 3′ signature of the sequence	No	No	Forward
5	1. The signature is seen in the forward strand of the cDNA 2. The cDNA has no polyA signals and no polyA tail 3. The signature is not the most 3′ signature of the sequence	No	No	Forward
11	1. The signature is seen in the reverse complementary strand of the cDNA 2. The reverse strand of the cDNA has a polyA signal and a polyA tail 3. The signature is the first one 5′ to the polyA signal and the polyA tail.	Yes	Yes	Reverse complement
12	1. The signature is seen in the reverse complement strand of the cDNA 2. The reverse complement strand of the cDNA has a polyA signal but no polyA tail 3. The signature is the first one 5′ to the polyA signal	Yes	No	Reverse complement
13	1. The signature is seen in the reverse complement strand of the cDNA 2. The reverse strand of the cDNA has a polyA tail but no polyA signals 3. The signature is the first one 5′ to the polyA tail	No	Yes	Reverse complement
14	1. The signature is seen in the reverse complement strand of the cDNA 2. The reverse complement strand of the cDNA has no polyA signals and no polyA tail 3. The signature is the most 3′ signature of the reverse complement strand of the cDNA sequence	No	No	Reverse complement

15	1. The signature is seen in the reverse complement strand of the cDNA 2. The reverse complement strand of the cDNA has no polyA signals and no polyA tails 3. The signature is not the most 3′ signature of the reverse complement strand of the cDNA sequence	No	No	Reverse Complement
20	1. The signature is seen in a cDNA of the “unknown” direction 2. The cDNA has no polyA signal and no polyA tail	No	No	Unknown
22	1. The signature is seen in a cDNA of the “unknown” direction 2. The cDNA has a polyA signal but no polyA tails 3. The signature is the first one 5′ to the polyA signal	Yes	No	Unknown
23	1. The signature is seen in a cDNA of the “unknown” direction 2. The cDNA has a polyA tail but no polyA signals 3. The signature is the first one 5′ to the polyA tail	No	Yes	Unknown
24	1. The signature is seen in a cDNA of the “unknown” direction 2. The cDNA has no polyA tail and no polyA signals 3. The signature is the first one 5′ to the polyA tail	No	No	Unknown
25	1. The signature is seen in a cDNA of the “unknown” direction 2. The cDNA has a polyA tail but no polyA signals 3. The signature is not the first one 5′ to the polyA tail	No	No	Unknown
1,000	The signature has only chromosome matches	NA	NA	Forward or reverse complement

(http://www.ncbi.nlm.nih.gov/sites/entrez?db=unigene). In fact, Unigene is our, and Lynx's, first choice in mapping tags to genes/transcripts. For those tags without any Unigene cluster matches, the matched ESTs are assigned. For those without any Unigene or EST matches, the matched genomic locations are recorded.

Recently, we have accumulated more prostate cancer MPSS data and have amassed a dataset for the prostate transcriptome with a total of over 22 million beads sequenced and 218,547 unique tags identified. In addition to tag comparison with Lynx's virtual database, we have performed a direct database search approach to annotate our MPSS datasets. We have employed a recently developed program miBLAST (Kim et al. 2005), in collaboration with Dr. J. M. Patel's group at the Univ. of Michigan. Our initial aim is to re-annotate those MPSS tags that are mapped to known transcripts (EST or cDNAs) in the GenBank database. In addition, we are curious as to whether allowing one mismatch will permit a broader and more comprehensive mapping to known transcripts. This is plausible, considering the high variability of the human genome and transcriptome (Goldstein and Cavalleri 2005; Levy et al. 2007). We would also like to analyze MPSS tags that map to intergenic regions. These regions may contain novel transcripts.

The three databases for miBLAST searches we used are: RefSeq – human.rna.fna (dated 03/26/2007, total 39,028 records); EST database: EST_human, (dated 03/08/2007, total 7,953,352 records); and the human Genome: human_genomic (reference assembly, complete sequence (NC_) were included. Total49 records.) (ftp://ftp.ncbi.nlm.nih.gov/RefSeq/H_sapiens/mRNA_Prot/ftp://ftp.ncbi.nlm.nih.gov/blast/db/FASTA/).

When we restricted true identifications to only those with perfect matches, we found that only about 20% (Table 6.2, 44, 485/218, 547) of the MPSS tags could be mapped to the human RefSeq databases. When we included the EST database, the chance of mapping a MPSS tag to a transcript increased dramatically from 20% to about 51% (Table 6.2, 110, 924/218, 547). Care should be taken, however, when interpreting the matching of a tag only to ESTs (i.e. those without RefSeq matches). RefSeq entries are usually curated and of high sequence quality, but genes represented only by ESTs could have a considerable amount of sequencing errors because ESTs are usually derived from "raw" sequencing data generated by only a single pass read.

Table 6.2 Summary of MPSS tag mapping by miBLAST with perfect matches

Databases	Total tags	Matched tags	Unmatched tags	Gene counts
Human_RefSeq	21,8547	44,485	1,74,062	18,013
EST_human	21,8547	1,10,924	1,07,623	18,437
Human_genomic	10,7152	52,367	54,785[a]	0

[a]There are 54,785 tags without any hit after searching against Refseq, EST, and genomic sequence database with perfect matches

Table 6.3 Summary of MPSS tag mapping by miBLAST allowing one mismatch

Databases	Total tags	Matched tags	Unmatched tags
Human_RefSeq	54,785	26,505	28,280
EST_human	54,785	53,138[a]	1,647
Human_genomic	1,621	1,541	80

[a]Almost 100% false positive rate assuming 1% sequence errors in ESTs

The remaining 50% of unmapped MPSS tags could represent uncharacterized portions of the genome, such as novel transcripts or new exons in known genes. About half of these tags did indeed map to the genome. To analyze those MPSS tags left (about 25% of the initial MPSS tags), we allowed one mismatch in mapping the tag to transcript. Any matches found this way should still carry weight, considering that the genome sequence of individuals, and thus the transcriptome sequence, may be more variable than previously thought (Levy et al. 2007). Because we discovered half of these one-mismatch tags mapped to RefSeq database (Table 6.3, 26, 505/54,785), it suggests that polymorphisms exist in the RefSeq or in the MPSS tags and normal polymorphisms should be considered when mapping MPSS tags (and, for that matter, future signal tags generated by next generation sequencing technologies). There have, in fact, been formal investigations studying the impact of polymorphisms (SNPs) on mapping MPSS tags to genes (Silva et al. 2004). Silva et al. observed that more than 8.6% of human genes have at least one alternative tag due to SNPs. They identified about 2,020 SNP-associated alternative tags from the human RefSeq databases and found that about 62% can be matched to MPSS or SAGE tags obtained from experiments. This is consistent with our analysis and suggests that it is important to consider normal polymorphisms when mapping tags to genes.

One should not, however, include one base mismatches in mapping MPSS tags to ESTs. Assuming 1% sequence error rate in the EST sequences, the high sequence error rate of ESTs would result in an enormous rate of false positives, which we estimated to be almost 99.7% false positives for a 20-bp tag allowing one mismatch as opposed to 18.2% false positives when only perfect matches were used. Interestingly, the error rate of an EST generated by large scale sequencing is position specific (i.e. the distance from the sequencing primer). The best regions are 150–200 nt from the sequencing primer and have an error rate of 0.23–0.36% but the worst regions (300–700 nt from the sequencing primer) can have error rates over 10% (Richterich 1998). This information has not yet been taken into account for mapping MPSS tags to ESTs and perhaps, in the future, a quality score can be assigned to the position within an EST, and much more reliable data will be retrieved. It is worth noting that most ESTs have sequence trace files available, so implementation of this algorithm will be possible.

There are some important complications to consider when analyzing the transcripts mapped by SAGE or MPSS tags. Because of the relatively small size of SAGE and MPSS tags, they can map to sequences common to more than one transcript (e.g. alternative splice variants or alternative polyadenylations from the same gene). It has been estimated that more than half of the human genes have multiple polyadenylation sites and thus have the potential to generate transcripts with different ending positions (Tian et al. 2005; Zhang et al. 2005). If MPSS is taken as an example, because the tags represent the region between the most 3′ DpnII site and the polyA tail, when a MPSS tag maps to a DpnII site that is 5′ to the most 3′ DpnII site, it could represent a true identification of an alternative polyadenylated transcript. Therefore, MPSS data have the potential to differentiate different polyadenylated transcripts from the same gene, and help us understand the effects of alternative polyadenylation on gene expression. Although this may appear to be more of an advantage than a complication, it is important to realize that these internally mapped tags could also be generated from incomplete digestion of the most 3′ DpnII sites, aberrant oligo d(T) priming to internal poly A stretches in the 3′ UTR during cDNA synthesis, and other experimental conditions. As such, there is a need for statistical or experimental development to address this hurdle.

Another complication arises from non-unique mapping: that is, SAGE or MPSS tags that map to multiple genes simultaneously. Lash et al. found that a considerable number of SAGE tags mapped to multiple genes (Lash et al. 2000), and we have also found that MPSS tags often map to multiple genes. Determining exactly which gene contributes to the observed counts becomes less reliable and more subject to inadvertent errors. To address this issue, in the schema Lynx used, if a tag matches to different unigene clusters, any unigene cluster against which the tag is annotated belonging to class 1–3 will take precedence. If a tag matches to multiple unigene clusters and the matching classes all belong to class1–3, the unigene cluster with the largest number of ESTs will be selected to represent the tag. In an attempt to circumvent the problems associated with non-unique mapping, Bianchetti et al. created the SAGETTARIUS algorithm for SAGE data analysis (Bianchetti et al. 2007). They developed four virtual tag databases from transcript databases of varying sequence quality: (1) high quality sequences from the RefSeq databases, which they named Cytoplasmic Ribosomal Transcript (CRT); (2) high quality sequences from individually cloned and verified cDNAs; (3) sequences from those generated by high-throughput cDNA (HTC) sequencing projects; and (4) low quality sequences from those represented by ESTs. SAGE tags were then mapped against these four virtual databases in descending order. In essence, Bianchetti et al. and others have employed methods to assign a "quality score" to their tag mapping data. Although their method undoubtedly helped produce highly reliable mapping data, it is important to realize that their sequence qualities between their assigned categories may overlap. Some of the clones from the 1980s and early 1990s may not have been of the highest quality, due to the limited sequencing technology available at the time (manual sequencing, low fidelity polymerase in the automatic sequencing, etc.). In contrast, some of the sequences from HTC projects are often high in quality. In essence, Bianchetti et al. and others have employed methods to assign a "quality score" to their tag mapping data.

6.3.2 Statistical Analysis of Signal Sequencing Data

The first step in comparative expression profiling of signal sequencing data is to estimate the normalized abundance of a transcript identified by a SAGE or MPSS tag. The normalization is performed by simply dividing the number of counts for a tag (transcript) by the sum of the observed counts for all the tags (transcripts) (i.e. transcript counts/total transcript counts). It is normally expressed as tags per million or transcripts per million (TPM).

Because signal sequencing based expression profiling is counting based, and because the counting is discrete, the data follows a discrete distribution, such as a binomial or Poisson distribution. SAGE data usually fits a binomial distribution. If $X1$ equals the observed counts for a transcript in sample 1 with a total count of $N1$, and $X2$ equals the observed counts for the same transcript in sample 2 with a total count of $N2$, then the comparison of transcript abundance between the two samples is: $p1 = x1/N1$ and $p2 = x2/N2$. To test the hypothesis $H0{:}\ p1 = p2$, we need to determine the level of significance. Because there are better statistical tests for Poisson or normal distributions, it is helpful to approximate the binomial distribution to the Poisson or the normal distribution. X is small compared to N in SAGE or MPSS data, so the binomial distribution can be reliably approximated to the Poisson distribution (in statistics, if $n \geq 100$ and $np \leq 10$, the Poisson distribution approximation of binomial distribution is dependable). Once approximated, statistics associated with the Poisson distribution can be applied for SAGE and MPSS data analysis, just as Madden et al. did for their SAGE data (Madden et al. 1997). In addition, if the N is very large (>1,00,000), such as that from a large SAGE or MPSS data set, one can approximate the binomial distribution to the normal distribution [in statistics, one can approximate a binomial distribution to a normal distribution when Np and $N(1-p)$ are greater than a certain assigned number (5 or 10 is usually taken in the statistics field)]. In this case, the formula $z = (p1 - p2)/sqrt(p(1 - p)(1/N1 + 1/N2))$ can be used to evaluate the significance for the sample comparison (Kal et al. 1999; Lin et al. 2005).

Various statistical tools are available on several SAGE analysis websites, such as SAGENET (Zhang et al. 1997)(http://www.sagenet.org/), SAGEmap (Lal et al. 1999) (http://www.ncbi.nlm.nih.gov/projects/SAGE/), SAGE Genie (Boon et al. 2002) (http://cgap.nci.nih.gov/SAGE), USAGE (van Kampen et al. 2000), eSAGE (Margulies and Innis 2000), WEBSAGE (Pylouster et al. 2005), SAGExplore (Norambuena et al. 2007) and IDEG6 (http://telethon.bio.unipd.it/bioinfo/IDEG6_form/) (Romualdi et al. 2003).

Numerous statistical tests have been proposed to analyze SAGE data in the above websites. In addition, other statistical methods have been proposed by others, such as the Z statistics (Kal et al. 1999), the Chi square test, Fisher's exact test, Audic and Claverie's Bayesian method (Audic and Claverie 1997), and Greller and Tobin's test (Greller and Tobin 1999). Man et al. (2000) systematically compared the Chi-square test, Fisher's exact test and Audic and Claverie's Bayesian method for comparative SAGE analysis and concluded that the Chi-square test had the best power (sensitivity) and robustness. Furthermore, some programs even offer multiple test correction such as Bonferroni Correction (e.g. in the IDEG6 website) and offer useful links to biological pathway and gene function category databases and websites. Since MPSS and SAGE data are very similar, many of these tools developed for SAGE analysis can be adapted to MPSS data analysis.

In our routine analysis, we used the Power SAGE method developed by Man et al. (Man et al. 2000). It should be noted that the *p* value is related to the abundance of transcripts. In pair-wise comparisons, we have noticed that a *P* value <0.001 can detect a 1.5-fold difference in highly abundant genes (e.g. 953 vs. 635 tpm in our LNCaP and CL1 data (Lin et al. 2005)), but a greater fold change is required to achieve that *P* value for low abundance transcripts (e.g. 26 tpm vs. 0 tpm).

Stolovitzky et al. at IBM (Stolovitzky et al. 2005) and Jared Roach from the Institute for Systems Biology developed a more sophisticated method to analyze MPSS data. They made use of the information contained in multiple replicated technical runs (multiple "flow cell") when the MPSS data was generated. For example, Lynx Therapeutics (now Illumina/Solexa Inc) usually runs at least two "two stepper" sequencing runs and two "four stepper" sequencing runs for a particular MPSS experiment. Stolovitzky and Roach modeled the internal noise in these replicates and devised a sophisticated statistical test based on probability theory for use in comparative MPSS data analysis.

6.4 Application of Signal Sequencing to Cancer Research

A check on the NCBI Gene Expression Omnibus (www.ncbi.nlm.nih.gov/geo) revealed that there are more than 958 SAGE datasets (samples) (http://www.ncbi.nlm.nih.gov/projects/geo/query/browse.cgi?mode=samples&filteron=8&filtervalue=4), and 180 MPSS datasets (samples) at the GEO database (http://www.ncbi.nlm.nih.gov/projects/geo/query/acc.cgi?acc=GPL1443). Many of these represent different types of cancer and cancer progression. The CGAP (the Cancer Genome Anatomy Project) either generated or collected 214 cancer SAGE datasets and their SAGE Genie website contains many web-based tools for SAGE analyses (http://cgap.nci.hih.gov/SAGE). Here, prostate and ovarian cancer will be used as an example to illustrate the application of the signal sequencing based expression profiling to cancer research.

6.4.1 Application of SAGE to Prostate and Ovarian Cancer Studies

Prostate cancer is the most common nondermatological cancer in the United States (Greenlee et al. 2000). Initially, its growth is androgen-dependent (AD); early-stage androgen deprivation therapies, including chemical and surgical castration, starve cancerous cells. Although such therapies initially result in tumor regression, they eventually fail when the prostate carcinomas inevitably become androgen-independent (AI) (Isaacs 1999; Debes and Tindall 2004). To improve the efficacy of prostate cancer therapy, it is necessary to understand the molecular mechanisms underlying the transition from androgen dependence to androgen independence.

There are about 15 SAGE datasets for prostate cancer in the NCI's CGAP database. Most of the prostate SAGE libraries were generated to identify androgen-regulated genes (ARGs) in prostate cancer cells (Waghray et al. 2001; Xu et al. 2001; Hu et al. 2002). A reanalysis of Waghray et al.'s data, using SAGE Genie's statistical analysis tool, revealed 420 differentially expressed tags between androgen starved and starved-then-stimulated LNCaP cells ($P < 0.05$, F = 1). When the same was done for Xu et al.'s data, 152 differentially expressed tags between androgen starved and starved-then-stimulated LNCaP cells were identified. The union of these two lists results in 545 tags that correspond to 504 genes. The list of genes include both novel androgen regulated genes and known androgen regulated genes, such as KLK3 and FMEPA1.

Untergasser et al. used SAGE to identify 157 up-regulated and 116 down-regulated mRNAs involved in senescence regulation in prostate epithelial cells (PrECs) (Untergasser et al. 2002). Walter-Yohrling et al. used SAGE to compare four invasive tumor cell lines (MDA-MB231, SKOV-3, A375, and MEL624) to four non-invasive tumor cell lines (LNCaP, DU145, PC3, and A549) to identify 47 differentially expressed genes that may be correlated with tumor invasion (Walter-Yohrling et al. 2003). There are also two SAGE libraries on the CGAP website obtained from microscope dissected normal and cancerous prostate tissues from Dr. G. Riggins's lab at Duke University Medical Center. An analysis using CGAP's online DGED analysis revealed that 81 tags ($P < 0.05$, F = 2) or 624 tags ($P < 0.05$, F = 1) were differentially expressed. Over-expressed genes in cancer cells included kallikrein 3 (KLK3), fatty acid synthase (FASN), SPINK1 (Serine peptidase inhibitor, Kazal type 1), and PLA2G2A (phospholipase A2, group IIA). Under-expressed genes included clusterin, transgelin, and actin gamma 2 (ACTG2).

In a similar manner Hough et al. used serial analysis of gene expression (SAGE) to compare gene expression profiles from various ovarian cell lines and tissues (i.e. normal ovarian surface epithelia cells and primary ovarian cancer cells) to identify up-regulated genes in ovarian cancer cells. They found that claudin 3, claudin 4, WAP four-disulfide core domain 2 (WFDC2, HE4), mucin-1, apolipoprotein E (ApoE), apolipoprotein J (ApoJ), and mesothelin were all up-regulated (Hough et al. 2000). Peters et al. applied SAGE to compare primary ovarian tumors to normal human ovarian surface epithelium (HOSE). They identified many genes over-expressed in ovarian tumors, some of which overlap with Hough et al.'s findings, including claudin 3 (CLDN3), WAP four-disulfide core domain 2 (WFDC2, HE4), folate receptor 1 (FLOR1), cyclin D1 and FLJ12988 (Peters et al. 2005).

High throughput studies like these are excellent at identifying potential cancer biomarkers and therapeutic drug targets. These candidates, however, need to be rigorously evaluated before any solid conclusions can be drawn. Recently, Hassan et al. developed a sandwich ELISA for the human mesothelin and found that 40 of 56 (71%) patients with mesothelioma and 14 of 21 (67%) patients with ovarian cancer have elevated serum mesothelin levels (Hassan et al. 2006). This is a solid start for SAGE's applicability to cancer serum biomarker research and, more generally, adds significant weight to the use of signal sequencing techniques for medically relevant research.

6.4.2 Application of MPSS to Prostate and Ovarian Cancer Studies

We have applied MPSS technology to study prostate cancer progression, in which prostate cancer cell lines, xenografts, and human clinical specimens were used to generate gene expression profiles (Lin et al. 2005, and unpublished data). As mentioned previously, understanding the cancer's progression to androgen independence (AI) is crucial to the advancement of prostate cancer therapies. The transition from AD to AI status likely results from multiple processes, including the activation of oncogenes, the inactivation of tumor suppressor genes, and other changes in key components of signalling pathways and gene regulatory networks(Isaacs 1999; Debes and Tindall 2004). Systems approaches to biology and disease are predicated on the identification of all the elements of a particular system, the delineation of their interactions, and their alteration in distinct disease states. Biological information consists of two types: the digital information of the genome (e.g. genes and their *cis*-controlling elements) and environmental cues or stimuli. Normal protein and gene regulatory networks may be perturbed by disease – through genetic and/or environmental perturbations – and understanding these differences lies at the heart of a systems approach to disease. Disease-perturbed networks initiate altered responses that bring about pathologic phenotypes, such as cancer cell invasiveness.

To map network perturbations in cancer initiation and progression, one must measure changes in expression levels of virtually all transcripts. Certain low-abundance transcripts, such as those encoding transcription factors and signal transducers, wield significant regulatory influences despite the fact they may be present in very low copy numbers. Differential display microarrays (Bussemakers et al. 1999) and cDNA microarrays (Chang et al. 1997; Vaarala et al. 2000) have been used to profile changes in gene expression during the AD to AI transition. However, those technologies have low detection sensitivities and consequently identify only a limited number of the more abundant mRNAs and miss many low-abundance mRNAs. Although transcriptome (mRNA levels) differences are easier to study than proteome (protein levels) differences, we all know cellular functions are primarily performed by proteins. Unfortunately, RNA expression profiling studies cannot address how the encoded proteins function biologically, and transcript abundance level does not necessarily correlate with protein level (Chen et al. 2002). We therefore attempted to complement our mRNA expression profiling with a more limited protein profile. We employed isotope-coded affinity tags (ICAT) coupled with tandem mass spectrometry (MS/MS) (Gygi et al. 1999). This approach enables differentially expressed genes to be mapped onto cellular networks and helps develop a systemic understanding of altered cellular states.

The LNCaP cell line is a widely used androgen-sensitive model for early-stage prostate cancer from which androgen-independent sublines have been generated (Chang et al. 1997; Patel et al. 2000; Vaarala et al. 2000). The cells of one such variant, CL-1, in contrast to their LNCaP progenitors, are highly tumorigenic, and exhibit invasive and metastatic characteristics in both intact and castrated mice (Patel et al. 2000; Tso et al. 2000). Thus CL-1 cells model late-stage AI prostate cancer. We conducted an MPSS analysis of about five million signatures for the androgen-dependent LNCaP cell line and its androgen-independent derivative CL1. Using very stringent *P* values (less than 0.001), we identified 2,088 MPSS signatures with significant differential expression (corresponding to 1,987 unique genes, as some genes have two or more MPSS signatures, due to alternative poly-adenylation sites). Of these, 1,011 signatures (965 genes) were differentially over-expressed in CL1 cells, and 1,077 signatures (1,022 genes) were differentially over-expressed in LNCaP cells.

We compared our LNCaP MPSS data against publicly available LNCaP SAGE data (NCBI SAGE database). We chose the SAGE library GSM724 (total SAGE tags sequenced: 22,721) (Lal et al. 1999), which is derived from LNCaP cells with an inactivated *PTEN* gene. Only 400 (about 20%) of our 1987 significantly differentially expressed genes ($P<0.001$) had any SAGE tag entry in GSM724. We identified many more differentially expressed genes, many of which were below the detection limits of SAGE analysis. These data illustrate the importance of deep sequence coverage in identifying state changes in transcripts expressed at low abundance levels.

We also performed functional classifications of genes differentially expressed between LNCaP and CL1 cells. The most interesting groups, categorized by function, are shown in Table 6.4. Furthermore, we compared the BioCarta and KEGG pathway databases (http://cgap.nci.nih.gov/Pathways/) with the MPSS data and identified 37 BioCarta and 14 KEGG pathways that are up-regulated in LNCaP cells when compared to CL1 cells, and identified 23 BioCarta and 22 KEGG pathways that appear to be down-regulated. Our database offers the first comprehensive digital transcriptome comparison between two distinct cellular states in the progression of prostate cancer and is a powerful tool for exploring the perturbed cellular pathways in the transition from AD to AI growth. It is worth mentioning that this dataset can be mapped to pathways and networks that are frequently updated.

When considering that gene regulatory networks are controlled by transcription factors, it logically follows to view transcription factors as key controllers of the progression of prostate cancer. It was extremely interesting to observe that, of 554 transcription factors expressed in LNCaP and CL1 cells, 112 showed significantly different mRNA expression levels between the cell lines ($P<0.001$) (Lin et al. 2005). It is also worth pointing out that, because many transcription factors are expressed at levels not reliably detected by DNA microarrays, this finding illustrates the power of

Table 6.4 Examples of differentially expressed genes and their functinonal classifications

Signatures	LNCaP(tpm)	CL1 (tpm)	Description	GenBank ID
Apoptosis related				
GATCAAATGTGTGGCCT	0	3,609	Lectin, galactoside-binding, soluble, 1 (galectin 1)	BC001693
GATCATAATGTTAACTA	0	14	Pleiomorphic adenoma gene-like 1 (PLAGL1)	NM_002656
GATCATCCAGAGGAGCT	0	16	Caspase 7, apoptosis-related cysteine protease	U40281
GATCGCGGTATTAAATC	0	15	Tumor necrosis factor receptor superfamily, member 12	U75380
GATCTCCTGTCCATCAG	0	24	Interleukin 1, beta	M15330
GATCCCCTTCAAGGACA	1	19	Nudix (nucleoside diphosphate linked moiety X)-type motif 1	NM_006024
GATCATTGCCATCACCA	51	278	EST, Highly similar to CUL2_HUMAN CULLIN HOMOLOG 2	AL832733
GATCTGAAAATTCTTGG	16	56	CASP8 and FADD-like apoptosis regulator	U97075
GATCCACCTTGGCCTCC	49	149	Tumor necrosis factor receptor superfamily, member 10b	NM_003842
GATCATGAATGACTGAC	118	257	Cytochrome c	BC009582
GATCAAGTCCTTTGTGA	299	102	Programmed cell death 8 (apoptosis-inducing factor)	H20713
GATCACCAAAACCTGAT	72	24	BCL2-like 13 (apoptosis facilitator)	BM904887
GATCAATCTGAACTATC	563	146	Apoptosis related protein APR-3 (APR-3)	NM_016085
GATCCCTCTGTACAGGC	83	13	unc-13-like (*C. elegans*) (UNC13), mRNA	NM_006377
GATCTGGTTGAAAATTG	1,006	49	CED-6 protein (CED-6), mRNA	NM_016315
GATCTCCCATGTTGGCT	86	4	CASP2 and RIPK1 domain containing adaptor with death domain	BC017042
GATCAGAAAATCCCTCT	27	1	DEAD/H (Asp-Glu-Ala-Asp/His) box polypeptide 20, 103 kDa	BC011556
GATCAAGGATGAAAGCT	50	3	Programmed cell death 2	D20426
GATCTGATTATTTACTT	1,227	321	Programmed cell death 5	NM_004708
GATCAAGTCCTTTGTGA	299	102	Programmed cell death 8 (apoptosis-inducing factor)	NM_004208
Cyclins				
GATCCTGTCAAAATAGT	2	47	MCT-1 protein (MCT-1), mRNA	NM_014060
GATCATTATATCATTGG	3	39	Cyclin-dependent kinase inhibitor 2B(CDKN2B)	NM_078487
GATCATCAGTCACCGAA	38	396	Cyclin-dependent kinase inhibitor 2A (p16)	BM054921
GATCGGGGGCGTAGCAT	5	43	Cyclin D1	NM_053056
GATCTACTCTGTATGGG	40	144	Cyclin fold protein 1	BG119256
GATCAGCACTCTACCAC	530	258	Cyclin B1	BM973693

GATCTGGTGTAGTATAT	210	77	Cyclin G2	BM984551
GATCAGTACACAATGAA	642	224	Cyclin G1	BC000196
GATCTCAGTTCTGCGTT	918	308	CDK2-associated protein 1 (CDK2AP1), mRNA	NM_004642
GATCCTGAGCTCCCTTT	2,490	650	Cyclin I	BC000420
GATCATGCAGTGACATA	15	1	KIAA1028 protein	AL122055
GATCTGTATGTGATTGG	28	1	Cyclin M3	AA489077
Kallikreins				
GATCCACACTGAGAGAG	841	0	KLK3	AA523902
GATCCAGAAATAAAGTC	385	0	KLK4	AA595489
GATCCTCCTATGTTGTT	314	0	KLK2	S39329
CD markers				
GATCAGAGAAGATGATA	0	810	CD213a2,interleukin 13 receptor, alpha 2	U70981
GATCCCTAGGTCTTGGG	23	161	CD213a1,interleukin 13 receptor, alpha 1	AW874023
GATCCACATCCTCTACA	0	63	CD33,CD33 antigen (gp67)	BC028152
GATCAATAATAATGAGG	0	151	CD44,CD44 antigen	AL832642
GATCCTTCAGCCTTCAG	0	35	CD73,5′-nucleotidase, ecto (CD73)	AI831695
GATCTGGAACCTCAGCC	1	50	CD49e,integrin, alpha 5	BC008786
GATCAGAGATGCACCAC	8	122	CD138,syndecan 1	BM974052
GATCAAAGGTTTAAAGT	38	189	CD166,activated leukocyte cell adhesion molecule	AL833702
GATCAGCTGTTTGTCAT	53	295	CD71,transferrin receptor (p90, CD71)	BC001188
GATCGGTGCGTTCTCCT	287	509	CD107a,lysosomal-associated membrane protein 1	AI521424
GATCTACAAAGGCCATG	161	681	CD29,integrin, beta 1	NM_002211
GATCATTTATTTTAAGC	56	0	CD10 (neutral endopeptidase, enkephalinase)	BQ013520
GATCAGTCTTTATTAAT	150	50	CD107b,lysosomal-associated membrane protein 2	AI459107
GATCTTGGCTGTATTTA	84	1,014	CD59 antigen p18-20	NM_000611
GATCTTGTGCTGTGCTA	408	234	CD9 antigen (p24)	NM_001769

(continued)

Table 6.4 (continued)

Signatures	LNCaP(tpm)	CL1 (tpm)	Description	GenBank ID
Transcription factors				
GATCAAATAACAAGTCT	0	62	Transcription factor BMAL2	BM854818
GATCTCTATGTTTACTT	0	27	Transcription factor BMAL2	BG163364
GATCCTGACACATAAGA	12	74	Transcription factor BMAL2	BF055294
GATCATTTTGTATTAAT	10	61	Transcription factor NRF	BC047878
GATCGTCTCATATTTGC	52	0	Transcriptional coactivator tubedown-100	NM_025085
GATCCCCCTCTTCAATG	0	31	Transcriptional co-activator with PDZ-binding motif	AJ299431
GATCAAATGCTATTGCA	1	55	Transcriptional regulator interacting with the PHS-bromodomain 2	AI126500
GATCTGTGACAGCAGCA	140	35	Transducer of ERBB2, 1	BC031406
GATCAAATCTGTACAGT	239	23	Transducer of ERBB2, 2	AA694240
Annexins and their ligands				
GATCCTGTGCAACAAGA	0	69	Annexin A10	BC007320
GATCTGTGGTGGCAATG	41	630	Annexin A11	AL576782
GATCAGAATCATGGTCT	0	1,079	Annexin A2	BC001388
GATCTCTTTGACTGCTG	210	860	Annexin A5	BC001429
GATCCAAAAACATCCTG	83	241	Annexin A6	AI566871
GATCAGAAGACTTTAAT	0	695	Annexin A1	BC001275
GATCAGGACACTTAGCA	0	2,949	S100 calcium binding protein A10 (annexin II ligand)	BC015973
Matrix metalloproteinase				
GATCATCACAGTTTGAG	0	38	Matrix metalloproteinase 10 (stromelysin 2)	BC002591
GATCCCAGAGAGCAGCT	0	108	Matrix metalloproteinase 1 (interstitial collagenase)	BC013118
GATCGGCCATCAAGGGA	0	25	Matrix metalloproteinase 13 (collagenase 3)	AI370581
GATCTGGACCAGAGACA	0	10	Matrix metalloproteinase 2 (gelatinase A)	BG332150

sensitive and comprehensive "deep" sequencing in signal sequence expression profiling. Yet another observation that demonstrates the endless opportunities for developing focused research from high throughput transcriptome data is our analysis of potentially secreted proteins from the LNCaP and CL1 data. From the 521 signatures identified, 460 genes potentially encoding secreted proteins (Lin et al. 2005). Among these, 259 and 201 genes, respectively, are over-expressed or under-expressed in the CL1 cells. It is exciting to consider the possibility that these changes in secreted protein expression will be detectable in the blood. If true, these alterations would reflect the underlying perturbed cancer networks and could be used to develop serum diagnostic tests that accurately monitor prostate cancer progression.

The extent of crosstalk among cellular signaling pathways has proven to be even more complicated and extensive than originally imagined. Therefore, rather than thinking signaling pathways as separate, discrete pathways, it is much more appropriate to think of them as interconnected to form expansive networks. Indeed, we have already found many of the perturbed pathways in the LNCaP and CL1 comparison to be interconnected into networks. The mapping of genes onto pathway networks will be an ongoing objective as more signaling information becomes available. Our transcriptome data should help delineate many of the complex relationships. In addition, by coupling this transcriptome data with the ICAT/MS/MS protein expression profiles comparing LNCaP and CL1 cells, we have taken a systems approach to prostate cancer by developing an integrative, systemic understanding of prostate cancer progression at the mRNA, protein and network levels.

We have also applied similar approaches to understand chemotherapy resistance in ovarian cancer, a major hurdle impeding the success of current therapies. Using MPSS technology and ICAT/MS/MS, we profiled the transcriptomes and proteomes of cisplatin sensitive (IGROV-1) and cisplatin resistant (IGROV-1/CP) ovarian cancer cell lines (Stewart et al. 2006). We obtained 3,422 signatures from the MPSS analysis that significantly differ between IGROV1 and IGROV1/CP cells ($P<0.001$) and a total of 1,117 proteins were identified and quantified by ICAT/MS/MS analysis. The relative protein expression of 121 of these varied between the two cell lines; 63 proteins were over-expressed in cisplatin sensitive and 58 were over-expressed in cisplatin resistant cells. Examples of proteins biologically relevant to cancer and at least fivefold over-expressed in resistant cells include the cell recognition molecule CASPR3 (13.3-fold), S100 protein family members (8.7-fold), claudin 4 junction adhesion molecule (7.2-fold) and the ATP binding protein, CDC42-binding protein (5.4-fold). Conversely, proteins at least fivefold over-expressed in the chemotherapy sensitive cells include hepatocyte growth factor inhibitor 1B (13.3-fold) and programmed cell death 6 protein (12.7-fold). As with the prostate cancer data, the protein expression profiles between cisplatin sensitive and resistant ovarian cancer cells were correlated to mRNA expression profiles. Again, these analyses are ongoing and as the pathways and networks are constantly updated, we are moving toward a more complete understanding, a systems biology understanding, of the chemotherapy response program at the mRNA, protein and network levels. This knowledge will undoubtedly aid in designing novel therapeutic approaches to overcome chemotherapy resistance

6.5 The Future of Signal Sequencing Based Expression Profiling

The development of next generation sequencing technology is revolutionizing genome sequencing as well as expression profiling. These more advanced sequencing methods and technologies in development include single molecule DNA sequencing, nanopore sequencing, sequencing by synthesis, sequencing by denaturation, pyrosequencing, sequence-by-ligation, and polony sequencing. Applied Biosystems, GE Healthcare, Helicos BioSciences, Illumina/Solexa, Intelligent Bio-Systems, Pacific Biosciences, Reveo, Roche/454 Life Sciences, VisiGen and several academic institutions (Harvard, Cornell, MIT, Caltech, Stanford) are heavily involved in sequencing technology development. We will focus our discussion below on a few of the more mature next generation sequencing technologies and their potential applications in signal sequencing based expression profiling.

The Pyrosequencing technology was developed by Ronaghi et al. at Stanford University (Ronaghi et al. 1996, 1998). The details are available on http://www.pyrosequencing.com/ and http://www.454.com/, but, in brief, it is based on the detection of pyrophosphate (PPi) released during DNA synthesis when inorganic PPi is released after nucleotide incorporation by DNA polymerase. The released PPi is then converted to ATP by ATP sulfurylase. A luciferase reporter enzyme uses the ATP to generate light, which is then detected by a charge coupled device (CCD) camera. The light signal is proportional to the number of nucleotides incorporated (e.g. A, TT, CCC etc.) and because the G, A, T, and C nucleotides are added stepwise in a sequencing cycle, the DNA sequences are easily derived. The discovery of apyrase has given pyrosequencing momentum for its implementation as a high throughput sequencing strategy. Apyrase continuously degrades unincorporated dNTPs and excess ATP, eliminating the need for a washing step between nucleotides in a pyrosequencing cycle (Ronaghi et al. 1998).

454 Life Sciences/Roche Diagnostics has commercial sequencing platforms based on the pyrosequencing technology (www.454.com). In this platform, DNAs are fractionated into 300–500 bp fragments and linkers are added to their 3′ and 5′ ends. Single stranded DNAs are isolated and captured on beads. The beads with DNAs are then emulsified in a 'water-in-oil' mixture with amplification reagents to create microreactors for emulsion PCR (emPCR). Finally, beads with amplified DNAs are loaded onto a PicoTiterPlate (allowing only one bead per well) for sequencing. One million or more DNA sequences can usually be obtained in a single pyrosequencing run, but the major advantage of the technology is its long sequence read length (>100 bases, approaching >500 bases), which enables extremely reliable mapping to the transcriptome or genome. However, of all the currently commercialized platforms, the 454 system has the greatest problems associated with sequencing long homopolymeric nucleotide regions. Bases are often inserted or deleted when these sequences are encountered. This could potentially create problems for tag-to-gene mapping in expression profiling, especially if only a short stretch is sequenced.

The application of pyrosequencing to expression profiling has been described in Agaton et al.'s pilot study, in which they sequenced the 3′ end fragments of cDNA libraries (Agaton et al. 2002). Recently, Bainbridge et al. applied pyrosequencing to sequence a cDNA library generated from LNCaP prostate cancer cells. They sequenced a total of 181,279 ESTs, which were mapped to about 10,000 gene loci in the human genome (Bainbridge et al. 2006). Unfortunately, this pilot study included only one library, so no comparative analyses could be performed.

Solexa/Illumina Inc. also has a next-generation sequencing platform – the Genome Analyzer system, which is also based on a SBS technology. They claim to have successfully solved the problem in sequencing long homopolymeric repeats with their novel reversible terminator chemistries. Furthermore, their Clonal Single Molecule Array technology increases the sequencing reaction signal by using "bridge amplification" in a flow cell to generate DNA clonal clusters (A detailed technology description is available at http://www.illumina.com/). This technology can potentially sequence an incredible number of DNA clones. There are eight channels in a flow cell, and each channel can generate >5–20 million DNA sequences. Recent improvements in its technology have enabled the reliable sequencing of about 36–100 bases, which should be sufficient for mapping them to the transcriptome or genome. Its throughput and sufficient read length give the technology immense potential for signal sequencing based expression profiling, as well as many other genomic analyses.

Applied Biosystems Inc. acquired Agencourt Personal Genomics who developed the SOLiD (Supported Oligo Ligation Detection) System for high throughput DNA sequencing, partly based on Dr. George Church's technology licensed from Harvard University. It is based on a sequencing-by-ligation method in combination with emulsion PCR (details at http://www.appliedbiosystems.com). In brief, DNA fragments are amplified by emulsion PCR, as we described earlier, and captured on beads. The beads are then loaded on a glass surface to form a random array for sequencing. Sequencing primers are added together with four oligo probes with different fluorescent labels (colors). Each oligo probe is eight-bases long and the first two bases are encoded. After hybridization, ligation and detection, sequences at the first and second position are determined. Repeating this process leads to the determination of dinucleotide sequences every five bases (e.g., at position 1,2,…6,7,…). After five cycles, the sequencing reaction is reset by denaturing the DNAs. Another sequencing primer that is offset by one base (n−1) is then added and the same sequence-by-ligation processes are repeated. The dinucleotide sequences at position 0,1, …5,6… are determined. Repeated sequencing cycles with sequencing primers that offset 2, 3, 4 bases (n−2, n−3, n−4) eventually lead to complete sequencing of the DNA. The read length is approaching 75 bases currently and each run can accommodate sequencing of >200–500 M reads with improvements number and length of reads planned.

The continued development of next generation sequencing technologies should soon make signal sequencing based expression profiling as cost effective as microarray based profiling. As described earlier, the information content and sensitivity provided by signal sequencing far surpasses that of microarray based technology. Indeed, recently, extensions of digital signal sequencing using the enormous sequencing capacity of these next generation sequencing strategies have allowed sequencing of the entire transcriptome including all sequences of all transcripts (RNA-seq or whole-transcriptome seq) without being limited to specific landmark sequences in each transcript (Cloonan et al. 2008; Cloonan and Grimmond 2008; Forrest and Carninci 2009). Such studies will ultimately allow discovery of all novel alternatively spliced transcripts, alternative transcriptional initiation and termination sites, allele-specific expression of genes, identification of fusion genes, etc. We should therefore expect to see most discovery-based studies using the powerful signal sequencing based expression profile technology. Microarrays, on the other hand, will still be useful for assessing the expression of a priori known genes across a great number of samples (e.g. gene-based diagnosis in clinical specimens).

References

Agaton C, Unneberg P, Sievertzon M, Holmberg A, Ehn M, Larsson M, Odeberg J, Uhlen M, Lundeberg J (2002) Gene expression analysis by signature pyrosequencing. Gene 289:31–39.

Audic S, Claverie JM (1997) The significance of digital gene expression profiles. Genome Res 7:986–995.

Bainbridge MN, Warren RL, Hirst M, Romanuik T, Zeng T, Go A, Delaney A, Griffith M, Hickenbotham M, Magrini V, Mardis ER, Sadar MD, Siddiqui AS, Marra MA, Jones SJ (2006) Analysis of the prostate cancer cell line LNCaP transcriptome using a sequencing-by-synthesis approach. BMC Genomics 7:246.

Bala P, Georgantas RW III, Sudhir D, Suresh M, Shanker K, Vrushabendra BM, Civin CI, Pandey A (2005) TAGmapper: a web-based tool for mapping SAGE tags. Gene 364: 123–129.

Bianchetti L, Wu Y, Guerin E, Plewniak F, Poch O (2007) SAGETTARIUS: a program to reduce the number of tags mapped to multiple transcripts and to plan SAGE sequencing stages. Nucleic Acids Res 35(18):e122.

Boon K, Osorio EC, Greenhut SF, Schaefer CF, Shoemaker J, Polyak K, Morin PJ, Buetow KH, Strausberg RL, De Souza SJ, Riggins GJ (2002) An anatomy of normal and malignant gene expression. Proc Natl Acad Sci U S A 99:11287–11292.

Brenner S, Johnson M, Bridgham J, Golda G, Lloyd DH, Johnson D, Luo S, McCurdy S, Foy M, Ewan M, Roth R, George D, Eletr S, Albrecht G, Vermaas E, Williams SR, Moon K, Burcham T, Pallas M, DuBridge RB, Kirchner J, Fearon K, Mao J, Corcoran K (2000a) Gene expression analysis by massively parallel signature sequencing (MPSS) on microbead arrays. Nat Biotechnol 18:630–634.

Brenner S, Williams SR, Vermaas EH, Storck T, Moon K, McCollum C, Mao JI, Luo S, Kirchner JJ, Eletr S, DuBridge RB, Burcham T, Albrecht G (2000b) In vitro cloning of complex mixtures of DNA on microbeads: physical separation of differentially expressed cDNAs. Proc Natl Acad Sci U S A 97:1665–1670.

Bussemakers MJ, van Bokhoven A, Verhaegh GW, Smit FP, Karthaus HF, Schalken JA, Debruyne FM, Ru N, Isaacs WB (1999) DD3: a new prostate-specific gene, highly overexpressed in prostate cancer. Cancer Res 59:5975–5979.

Chang GT, Blok LJ, Steenbeek M, Veldscholte J, van Weerden WM, van Steenbrugge GJ, Brinkmann AO (1997) Differentially expressed genes in androgen-dependent and -independent prostate carcinomas. Cancer Res 57:4075–4081.

Chen G, Gharib TG, Huang CC, Taylor JM, Misek DE, Kardia SL, Giordano TJ, Iannettoni MD, Orringer MB, Hanash SM, Beer DG (2002) Discordant protein and mRNA expression in lung adenocarcinomas. Mol Cell Proteomics 1:304–313.

Cloonan N, Grimmond SM (2008) Transcriptome content and dynamics at single-nucleotide resolution. Genome Biol 9:234.

Cloonan N, Forrest AR, Kolle G, Gardiner BB, Faulkner GJ, Brown MK, Taylor DF, Steptoe AL, Wani S, Bethel G, Robertson AJ, Perkins AC, Bruce SJ, Lee CC, Ranade SS, Peckham HE, Manning JM, McKernan KJ, Grimmond SM (2008) Stem cell transcriptome profiling via massive-scale mRNA sequencing. Nat Methods 5:613–619.

Debes JD, Tindall DJ (2004) Mechanisms of androgen-refractory prostate cancer. N Engl J Med 351:1488–1490.

Forrest AR, Carninci P (2009) Whole genome transcriptome analysis. RNA Biol 6:107–112.

Goldstein DB, Cavalleri GL (2005) Genomics: understanding human diversity. Nature 437:1241–1242.

Greenlee RT, Murray T, Bolden S, Wingo PA (2000) Cancer statistics, 2000. CA Cancer J Clin 50:7–33.

Greller LD, Tobin FL (1999) Detecting selective expression of genes and proteins. Genome Res 9:282–296.

Gygi SP, Rist B, Gerber SA, Turecek F, Gelb MH, Aebersold R (1999) Quantitative analysis of complex protein mixtures using isotope-coded affinity tags. Nat Biotechnol 17:994–999.

Hassan R, Remaley AT, Sampson ML, Zhang J, Cox DD, Pingpank J, Alexander R, Willingham M, Pastan I, Onda M (2006) Detection and quantitation of serum mesothelin, a tumor marker for patients with mesothelioma and ovarian cancer. Clin Cancer Res 12:447–453.

Hough CD, Sherman-Baust CA, Pizer ES, Montz FJ, Im DD, Rosenshein NB, Cho KR, Riggins GJ, Morin PJ (2000) Large-scale serial analysis of gene expression reveals genes differentially expressed in ovarian cancer. Cancer Res 60:6281–6287.

Hu YF, Ren ZY, Li YF, Sun HX, Chang YS, Su CB, Wang RZ, Zuo J, Fang FD (2002) Serial analysis of gene expression in the pituitary adenomas and para-tumor normal pituitary tissues. Zhongguo Yi Xue Ke Xue Yuan Xue Bao 24:611–615.

Isaacs JT (1999) The biology of hormone refractory prostate cancer. Why does it develop? Urol Clin North Am 26:263–273.

Kal AJ, van Zonneveld AJ, Benes V, van den Berg M, Koerkamp MG, Albermann K, Strack N, Ruijter JM, Richter A, Dujon B, Ansorge W, Tabak HF (1999) Dynamics of gene expression revealed by comparison of serial analysis of gene expression transcript profiles from yeast grown on two different carbon sources. Mol Biol Cell 10:1859–1872.

Kim YJ, Boyd A, Athey BD, Patel JM (2005) miBLAST: scalable evaluation of a batch of nucleotide sequence queries with BLAST. Nucleic Acids Res 33:4335–4344.

Lal A, Lash AE, Altschul SF, Velculescu V, Zhang L, McLendon RE, Marra MA, Prange C, Morin PJ, Polyak K, Papadopoulos N, Vogelstein B, Kinzler KW, Strausberg RL, Riggins GJ (1999) A public database for gene expression in human cancers. Cancer Res 59:5403–5407.

Lash AE, Tolstoshev CM, Wagner L, Schuler GD, Strausberg RL, Riggins GJ, Altschul SF (2000) SAGEmap: a public gene expression resource. Genome Res 10:1051–1060.

Levy S, Sutton G, Ng PC, Feuk L, Halpern AL, Walenz BP, Axelrod N, Huang J, Kirkness EF, Denisov G, Lin Y, Macdonald JR, Pang AW, Shago M, Stockwell TB, Tsiamouri A, Bafna V, Bansal V, Kravitz SA, Busam DA, Beeson KY, McIntosh TC, Remington KA, Abril JF, Gill J, Borman J, Rogers YH, Frazier ME, Scherer SW, Strausberg RL, Venter JC (2007) The diploid genome sequence of an individual human. PLoS Biol 5:e254.

Lin B, White JT, Lu W, Xie T, Utleg AG, Yan X, Yi EC, Shannon P, Khrebtukova I, Lange PH, Goodlett DR, Zhou D, Vasicek TJ, Hood L (2005) Evidence for the presence of disease-perturbed networks in prostate cancer cells by genomic and proteomic analyses: a systems approach to disease. Cancer Res 65:3081–3091.

Madden SL, Galella EA, Zhu J, Bertelsen AH, Beaudry GA (1997) SAGE transcript profiles for p53-dependent growth regulation. Oncogene 15:1079–1085.

Man MZ, Wang X, Wang Y (2000) POWER_SAGE: comparing statistical tests for SAGE experiments. Bioinformatics 16:953–959.

Margulies EH, Innis JW (2000) eSAGE: managing and analysing data generated with serial analysis of gene expression (SAGE). Bioinformatics 16:650–651.

Matsumura H, Reich S, Ito A, Saitoh H, Kamoun S, Winter P, Kahl G, Reuter M, Kruger DH, Terauchi R (2003) Gene expression analysis of plant host–pathogen interactions by SuperSAGE. Proc Natl Acad Sci U S A 100:15718–15723.

Norambuena T, Malig R, Melo F (2007) SAGExplore: a web server for unambiguous tag mapping in serial analysis of gene expression oriented to gene discovery and annotation. Nucleic Acids Res 35:W163–168.

Patel BJ, Pantuck AJ, Zisman A, Tsui KH, Paik SH, Caliliw R, Sheriff S, Wu L, deKernion JB, Tso CL, Belldegrun AS (2000) CL1-GFP: an androgen independent metastatic tumor model for prostate cancer. J Urol 164:1420–1425.

Peters DG, Kudla DM, Deloia JA, Chu TJ, Fairfull L, Edwards RP, Ferrell RE (2005) Comparative gene expression analysis of ovarian carcinoma and normal ovarian epithelium by serial analysis of gene expression. Cancer Epidemiol Biomarkers Prev 14:1717–1723.

Porter D, Yao J, Polyak K (2006) SAGE and related approaches for cancer target identification. Drug Discov Today 11:110–118.

Pylouster J, Senamaud-Beaufort C, Saison-Behmoaras TE (2005) WEBSAGE: a web tool for visual analysis of differentially expressed human SAGE tags. Nucleic Acids Res 33:W693–695.

Reinartz J, Bruyns E, Lin JZ, Burcham T, Brenner S, Bowen B, Kramer M, Woychik R (2002) Massively parallel signature sequencing (MPSS) as a tool for in-depth quantitative gene expression profiling in all organisms. Brief Funct Genomic Proteomic 1:95–104.

Richterich P (1998) Estimation of errors in "raw" DNA sequences: a validation study. Genome Res 8:251–259.

Romualdi C, Bortoluzzi S, D'Alessi F, Danieli GA (2003) IDEG6: a web tool for detection of differentially expressed genes in multiple tag sampling experiments. Physiol Genomics 12:159–162.

Ronaghi M, Karamohamed S, Pettersson B, Uhlen M, Nyren P (1996) Real-time DNA sequencing using detection of pyrophosphate release. Anal Biochem 242:84–89.

Ronaghi M, Uhlen M, Nyren P (1998) A sequencing method based on real-time pyrophosphate. Science 281(363):365.

Saha S, Sparks AB, Rago C, Akmaev V, Wang CJ, Vogelstein B, Kinzler KW, Velculescu VE (2002) Using the transcriptome to annotate the genome. Nat Biotechnol 20:508–512.

Silva AP, De Souza JE, Galante PA, Riggins GJ, De Souza SJ, Camargo AA (2004) The impact of SNPs on the interpretation of SAGE and MPSS experimental data. Nucleic Acids Res 32:6104–6110.

Stewart JJ, White JT, Yan X, Collins S, Drescher CW, Urban ND, Hood L, Lin B (2006) Proteins associated with Cisplatin resistance in ovarian cancer cells identified by quantitative proteomic technology and integrated with mRNA expression levels. Mol Cell Proteomics 5:433–443.

Stolovitzky GA, Kundaje A, Held GA, Duggar KH, Haudenschild CD, Zhou D, Vasicek TJ, Smith KD, Aderem A, Roach JC (2005) Statistical analysis of MPSS measurements: application to the study of LPS-activated macrophage gene expression. Proc Natl Acad Sci U S A 102:1402–1407.

Tian B, Hu J, Zhang H, Lutz CS (2005) A large-scale analysis of mRNA polyadenylation of human and mouse genes. Nucleic Acids Res 33:201–212.

Tso CL, McBride WH, Sun J, Patel B, Tsui KH, Paik SH, Gitlitz B, Caliliw R, van Ophoven A, Wu L, deKernion J, Belldegrun A (2000) Androgen deprivation induces selective outgrowth of aggressive hormone- refractory prostate cancer clones expressing distinct cellular and molecular properties not present in parental androgen-dependent cancer cells. Cancer J Sci Am 6:220–233.

Untergasser G, Koch HB, Menssen A, Hermeking H (2002) Characterization of epithelial senescence by serial analysis of gene expression: identification of genes potentially involved in prostate cancer. Cancer Res 62:6255–6262.

Vaarala MH, Porvari K, Kyllonen A, Vihko P (2000) Differentially expressed genes in two LNCaP prostate cancer cell lines reflecting changes during prostate cancer progression. Lab Invest 80:1259–1268.

van Kampen AH, van Schaik BD, Pauws E, Michiels EM, Ruijter JM, Caron HN, Versteeg R, Heisterkamp SH, Leunissen JA, Baas F, van der Mee M (2000) USAGE: a web-based approach towards the analysis of SAGE data. Serial analysis of gene expression. Bioinformatics 16:899–905.

Velculescu VE, Zhang L, Vogelstein B, Kinzler KW (1995) Serial analysis of gene expression. Science 270:484–487.

Waghray A, Feroze F, Schober MS, Yao F, Wood C, Puravs E, Krause M, Hanash S, Chen YQ (2001) Identification of androgen-regulated genes in the prostate cancer cell line LNCaP by serial analysis of gene expression and proteomic analysis. Proteomics 1:1327–1338.

Walter-Yohrling J, Cao X, Callahan M, Weber W, Morgenbesser S, Madden SL, Wang C, Teicher BA (2003) Identification of genes expressed in malignant cells that promote invasion. Cancer Res 63:8939–8947.

Xu LL, Su YP, Labiche R, Segawa T, Shanmugam N, McLeod DG, Moul JW, Srivastava S (2001) Quantitative expression profile of androgen-regulated genes in prostate cancer cells and identification of prostate-specific genes. Int J Cancer 92:322–328.

Zhang L, Zhou W, Velculescu VE, Kern SE, Hruban RH, Hamilton SR, Vogelstein B, Kinzler KW (1997) Gene expression profiles in normal and cancer cells. Science 276:1268–1272.

Zhang H, Lee JY, Tian B (2005) Biased alternative polyadenylation in human tissues. Genome Biol 6:R100.

Chapter 7
Mass Spectrometry Based Proteomics in Cancer Research

Mohamad A. Abbani, Parag Mallick, and Maryann S. Vogelsang

Abstract Proteomics has become an important component of biological and clinical research. Numerous proteomics methods have been developed to identify and quantify the proteins present in biological and clinical samples (Gerber et al., Proc Natl Acad Sci U S A 100:6940–6945, 2003; Ong et al., Methods 29:124–130, 2003). Differences among cell types or treatment groups have been used to identify cellular functions and pathways affected by disease or perturbations (Wright et al., Genome Biol 5:R4, 2003; Durr et al., Nat Biotechnol 22:985–992, 2004), new components and changes in the composition of protein complexes and organelles (Andersen et al., Nature 426:570–574, 2003; Blagoev et al., Nat Biotechnol 21:315–318, 2003; Ranish et al., Nat Genet 36:707–713, 2004), and putative disease biomarkers (Marko-Varga et al., J Proteome Res 4:1200–1212, 2005). Despite widespread success, the application of these approaches to discovery of relevant protein markers from clinical samples has been hampered by sample complexity and variability. To begin to broach this challenge, complex experimental protocols for enrichment, separation, and quantification have been developed for selective or comprehensive proteome analysis. In this chapter, we describe techniques for enrichment, separation, quantification, fundamentals of mass spectrometry, and the computational analysis of data generated by these processes within the context of using these approaches for asking and answering biologically and clinically important questions.

P. Mallick (✉)
Center for Applied Molecular Medicine, University of Southern California, Los Angeles, CA, USA
and
Department of Chemistry and Biochemistry, University of California, Los Angeles, CA, USA
e-mail: parag@ucla.edu

S. Yegnasubramanian and W.B. Isaacs (eds.), *Modern Molecular Biology*, Applied Bioinformatics and Biostatistics in Cancer Research,
DOI 10.1007/978-0-387-69745-1_7,

7.1 Introduction

Proteomics has become an important component of biological and clinical research. Numerous proteomics methods have been developed to identify and quantify the proteins present in biological and clinical samples (Gerber et al. 2003; Ong et al. 2003). Differences among cell types or treatment groups have been used to identify cellular functions and pathways affected by disease or perturbations (Wright et al. 2003; Durr et al. 2004), new components and changes in the composition of protein complexes and organelles (Andersen et al. 2003; Blagoev et al. 2003; Ranish et al. 2004), and putative disease biomarkers (Marko-Varga et al. 2005). Despite widespread success, the application of these approaches to discovery of relevant protein markers from clinical samples has been hampered by sample complexity and variability. To begin to broach this challenge, complex experimental protocols for enrichment, separation, and quantification have been developed for selective or comprehensive proteome analysis. In this chapter, we describe techniques for enrichment, separation, quantification, fundamentals of mass spectrometry, and the computational analysis of data generated by these processes within the context of using these approaches for asking and answering biologically and clinically important questions.

Generally, we can define proteomics as the systematic study of the many and diverse properties of the proteins in a system with the aim of providing detailed descriptions of the structure, function, and control of biological systems in health and disease. Advances in methods and technologies have catalyzed an expansion of the scope of biological studies from the reductionist biochemical analysis of single proteins to proteome-wide measurements. Proteomics, like other high-throughput "discovery" approaches, such as genomic sequencing, microarray analysis, and metabolite profiling, has been catalyzed by mapping and sequencing of the complete genomes of many species. Through the sequencing of a genome, we are able to generate an approximate estimate the scale of the proteome. However, it is critical to recognize that the proteome is fundamentally different in nature than the genome. The proteome is significantly more complex than the genome as multiple protein isoforms can be synthesized from a single gene and as proteins have greater chemical diversity (due to post-translational modifications (glycosylation, methylation, proteolytic cleavage, etc.). Proteins are found over a wide dynamic range of concentration (10^8 per cell to 10^{12} in biological fluids) (Anderson and Anderson 2002). In addition, the genome is relatively static over the lifespan of an organism, whereas the proteome is highly dynamic, changing on rapid timescales in response to both environmental and chemical perturbations to the system.

Over the years, mass spectrometry has become the method of choice for proteomic analysis. This technology has enabled us to characterize complex mixtures and probe for detailed information about individual proteins (e.g., covalent structures and post-translational modifications). A variety of proteomics applications exist, including the study of protein–protein interactions via affinity-based isolations on a small and proteome-wide scale, the mapping of numerous organelles, the concurrent description of the genome and proteome of small organisms (e.g., tuberculosis and malaria), and the generation of differential quantitative protein profiles of diverse species.

The ability of mass spectrometry to identify and, increasingly, to precisely quantify thousands of proteins from complex samples can be expected to broadly impact biology and medicine. Furthermore, proteomics has become an integral part in the emerging field of systems biology.

The essence of systems biology approaches pre-supposes that for any given system, the space of possible biomolecules and their organization into pathways and processes is large but finite. Consequently, the biological systems operating in an organism can be described comprehensively if a sufficient density of observations on all of the elements that constitute the system can be obtained. Proteomics is a particularly rich source of biological information because proteins are involved in almost all biological activities and they also have diverse properties, which collectively contribute greatly to our understanding of biological systems.

A standard proteomics process has three main components: (1) sample preparation, (2) mass spectrometric analysis, (3) data analysis and interpretation. In this chapter we initially describe these three aspects of the process. We conclude by describing specific examples of proteomics applications, including protein-protein interaction characterization, protein post-translational modification characterization, and quantitative differential analyses (both unbiased and targeted). In the interest of space we do not specifically describe the cannon of techniques for extraction of a sample from a patient or from in vitro experiments. Instead, we focus on the techniques for sample generation using metabolic labeling, used in quantitative proteomics, and on the downstream analysis once a protein mixture has been obtained.

7.2 Sample Preparation

7.2.1 Proteome Analysis Challenged by the Large Concentration Range

Evaluation of the human proteome provides opportunities to improve disease diagnosis and therapeutic monitoring. However, before these goals are realized, challenges must be overcome. For example, the human plasma proteome has inherent properties that complicate analysis by mass spectrometry. It contains a densely concentrated number of proteins and has a large dynamic concentration range of proteins that exceeds 10 orders of magnitude (Anderson and Anderson 2002). This large dynamic range is further complicated as a large fraction of the proteins are albumin (55%) and glycoproteins. In addition, it contains subsets of proteins from other tissues. Regardless of these challenges, human plasma can be analyzed. Advantages include abundance and ease of collection in the clinic.

The massive complexity of the plasma proteome requires that it be divided by fractionation into manageable smaller parts prior to analysis by currently available analytical methods. Although the physiochemical properties of proteins offer scientists a rich basis for many separation techniques (Table 7.1), protein solubility is considered the bottleneck problem in fractionation. To solubilize proteins, detergents

Table 7.1 A summary of fractionation techniques

Method		Properties based separations	Application	Separation	% Reproducibility
Chromatography					80–90
	Gel filtration	Size	Fractionation	Poor	
	Ion exchange	Electrostatic	Fractionation	Moderate	
	Chromatofocusing	Isoelectric point	Fractionation	Good	
	RPLC	Hydrophobicity	Fractionation	Moderate	
	Metal chelate	Electrostatic	Enrichment	Good	
	Affinity	Structure/ligand binding capacity	Fractionation/enrichment	Good	
Electrophoresis		Size and P*I*	Fractionation	Good	40–60

are often used; the choice of detergent often puts limits on the types of fractionation that can be used. Sample recovery and reproducibility are of great importance. Several methods of protein fractionation are discussed in the following section. The most common paradigm in proteomics, "bottom-up" proteomics, operates on peptide fragments of proteins. These peptides are typically generated by digestion of a protein mixture with a protease (e.g., trypsin). The fractionation methods below can be applied either on protein mixture prior to digestion, or on peptide mixtures following digestion.

7.2.2 *Fractionation Using Chromatographic Techniques*

In general, for chromatographic separations, proteins interact with a solid phase and are then released into a liquid phase and recovered (Fig. 7.1). In general, chromatographic

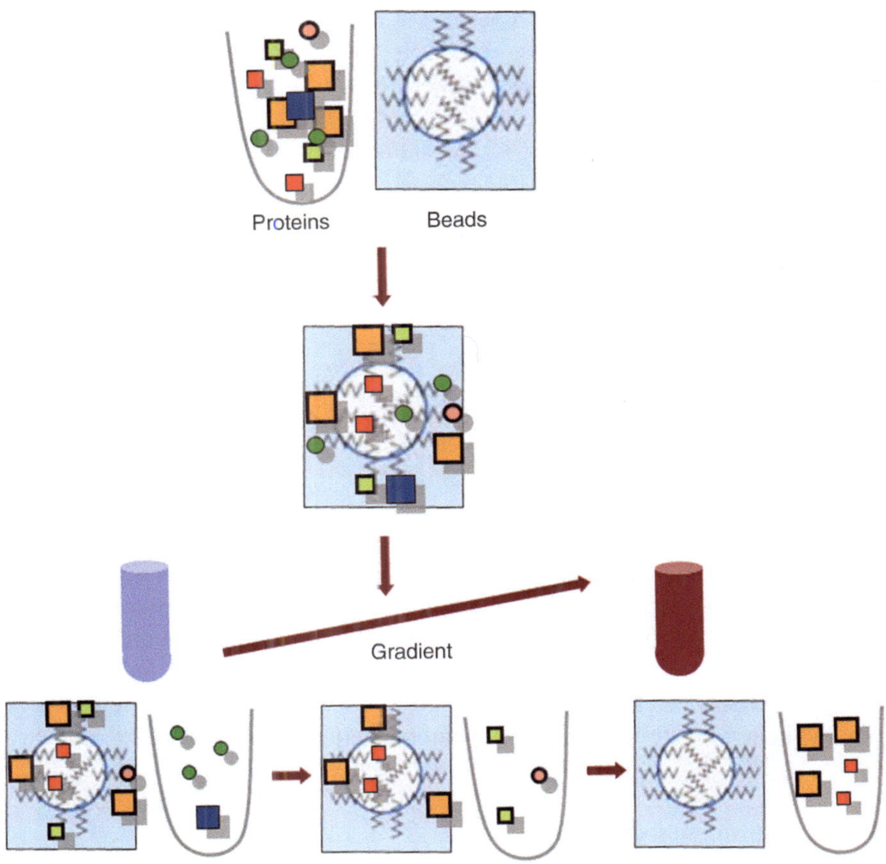

Fig. 7.1 Schematic diagram of the general process of chromatography. First, proteins are loaded on the column, where they interact with the solid phase. Second, proteins are released gradually by a gradient to be recovered into the liquid phase separated into different fractions

methods offer reproducible separation with high recovery of proteins as compared to electrophoresis techniques. Based on protein separation parameters, chromatographic techniques are categorized as follows:

7.2.2.1 Gel Filtration

The gel filtration process separates proteins based on size. In this method, the proteins travel through a column with a solid phase made of permeable beads. Large proteins cannot enter the pores of the beads; they pass around the beads and travel a shorter distance than smaller proteins to elute first. Smaller proteins penetrate the pores and pass through the beads, resulting in retardation of migration through the column. In gel filtration, proteins elute in the order of decreasing size. Although this technique offers good recovery and reproducibility, it suffers from low resolution as compared to other procedures.

7.2.2.2 Ion Exchange Chromatography

On ion exchange resin, proteins are separated based on their electrical charge. Proteins bind to the solid phase ion exchange resin through electrostatic interactions between charges on the proteins and of the opposite charges on the resin. The mobile phase used for loading the proteins onto the column is electrically neutral, so no interference with protein binding is introduced. Proteins are eluted as the ionic strength of the mobile phase is increased; this increases the competition for binding of proteins to the solid phase and proteins are eluted in order of affinity for the ionic resin. Less commonly, the pH of the solution can be adjusted such that the charge on either the proteins or the solid phase is altered to dislodge the analytes. Although the ion exchange procedure is limited by protein solubility, it offers higher resolution separation than gel filtration. For instance, charged detergents cannot be used in binding solution when applying this technique, but neutral ones offer a great advantage in solubilizing proteins. In addition, native or denatured proteins can be separated in the presence or absence of non-interfering detergents. A limitation of this technique is that a multicharge difference is required between species for good resolution, rendering it unpractical for post-translational modification studies. However, it is still considered a very powerful and useful tool for protein fractionation. This approach has been refined for efficient peptide separation by the use of strong ion exchange chromatography (Burke et al. 1989).

7.2.2.3 Chromatofocusing

Another valuable technique is chromatofocusing, a variant of ion exchange chromatography. In this technique, fractionation is based on the protein's isoelectric point (P*I*). The binding and elution of proteins is controlled solely by the pH of the mobile phase. In anion chromatofocusing, binding is achieved at high pH and elution is accomplished by introducing a pH gradient. As the pH on the column is decreased, the positive charge on the proteins becomes more pronounced, the negative charge of the column becomes weaker and proteins are dislodged. The opposite process is followed for cation chromatofocusing. As the proteins are dissociated from the top of the column due to the change in pH, they re-bind at a lower region where the pH is still favorable. This process is repeated until the proteins reach the bottom of the column and elute in a very concentrated volume; this leads to greater resolution than is obtained on ion exchange. Like ion-exchange, protein solubility poses a challenge to the chromatofocusing process. However, unlike ion-exchange, this technique can be used to distinguish post-translational modifications to proteins, since even a single modification can lead to a different P*I*.

7.2.2.4 Reversed Phase Chromatography

Hydrophobicity is another parameter used in protein fractionation. The most common technique in this class is high-performance liquid chromatography (HPLC), a special type of reversed phase chromatography. In general, a non-polar stationary phase and an aqueous, moderately polar mobile phase are employed. Proteins that are non-polar have a longer retention time, whereas polar molecules elute quickly. By increasing the non-polar character of the mobile phase, adsorbed proteins are eluted. For proteomic studies, the proteins from whole cell lysates or biological fluids are denatured prior to fractionation.

7.2.2.5 Metal Chelate Chromatography

Proteins can also be fractionated by metal chelate chromatography, in which a metal ion is affixed to the solid phase via an immobilized iminodiacetic acid resin. All metal binding compounds are attracted to this surface; elution is carried out using solutions containing competing molecules such as immidazole or by varying the pH. For proteomic studies, this method is commonly applied for enrichment. For example, immobilized metal affinity chromatography (IMAC) can be used to extract phospho-proteins/peptides from complex mixtures (Porath 1992). In this approach, the phospho-peptides are captured by a trivalent cation complex and then eluted by either high pH or a phosphate buffer. To minimize non-specific binding from peptides rich in carboxylate groups, tryptic peptides are converted to methyl esters using methanolic HCl prior to enrichment (Ficarro et al. 2002).

7.2.2.6 Affinity Chromatography

A variety of proteins can be selectively captured by a matrix immobilized ligand. The ligand–protein complex is then destabilized by salts or by competition through another ligand-binding entity. Although, this technique has been applied successfully to fractionation, it is used also for complex sample depletion of abundant proteins to reduce mixture complexity. For example, this strategy has been utilized in proteomic studies of serum/plasma and other body fluids to enhance the detection of low abundance proteins and achieve broader proteome coverage. In human plasma, there are 22 most abundant proteins responsible for ~99% of the total protein mass and these proteins mask the detection of hundreds of thousands of other proteins (Anderson and Anderson 2002).

In addition, affinity chromatography is used for enrichment of glycoproteins and glycopeptides. One method of glyco-capture is the use of lectin affinity chromatography (Cummings and Kornfeld 1982; Hirabayashi 2004). In this method, *N*-glycoproteins are captured through their binding to immobilized concanavalin A (Con A) (Bunkenborg et al. 2004; Fan et al. 2004). This method has been refined by combining Con A with wheat germ agglutinin (WGA) for the capture *O*-glycopeptides/proteins (Bunkenborg et al. 2004; Yang and Hancock 2004). Enrichment of glycoproteins or peptides can be achieved by other approaches discussed above. For example, size exclusion chromatography is utilized since most tryptic glycopeptides in a complex mixture have a relatively high mass (Alvarez-Manilla et al. 2006). Hydrophilicity of the glycan moiety is also utilized for enrichment through the hydrophilic interaction with a carbohydrate-based matrices (Wada et al. 2004).

7.2.3 Fractionation by Gel Electrophoresis

This type of protein separation exploits the size and charge or isoelectric point of proteins as parameters for separation. The proteins are driven through the gel matrix by an electric current. The most common method that uses size and charge as parameters for separation is the sodium dodecyl sulfate (SDS) electrophoresis. In general, the matrix is made up of acrylamide crosslinked to produce differently sized porous networks. Initially, proteins are uniformly charged with SDS (1.4 g SDS/g protein) (Reynolds and Tanford 1970). This SDS–protein coupling prevents aggregation, as all proteins are highly negatively charged, and produces a uniform charge to mass ratio for all proteins such that separation occurs based solely on size. Denatured proteins are loaded in wells in the gel and when the electric current is on they move toward the anode. Gels are usually sectioned into two steps to achieve better resolution. In the first step (stacking), proteins are concentrated by an isotachophoresis process. In the second step, proteins are separated based on their ability to navigate the matrix and recovered by excision for MS analysis. One disadvantage of SDS electrophoresis is its inability to separate post-translationally modified proteins. Another drawback is that some proteins, including glycoproteins and very basic proteins, have poor detergent binding. Also, the high pH conditions of electrophoresis may remove or introduce some post-translational modifications. BAC-SDS was developed by Macfarlane et al. to avoid alteration of post-translational modifications. The first dimension of BAC-SDS is a cationic detergent based electrophoresis and the second dimension is SDS-based.

Isoelectric focusing (IEF) uses the isoelectric point of proteins as a separation parameter. In IEF, separation is achieved by placing proteins in a pH gradient and driving movement by an electric current. Once proteins reach their isoelectric point zone, their charge becomes zero and they cease movement. Similar to chromatofocusing discussed above, this technique provides high resolution, even when proteins differ slightly (as little as 0.1 pH unit in their P*I*). This makes IEF a very useful tool for separating proteins and their post-translationally modified forms. Isoelectric focusing is widely used in proteomics in a stepwise combination with SDS-PAGE resulting in a two-dimensional electrophoresis (2DE). First, proteins are fractionated using IEF and next are resolved on SDS-PAGE gels. Although gel-based separation techniques have been widely used, they suffer major drawbacks, including poor reproducibility and low protein recovery.

Given the specificity and restrictions of each of the above mentioned methods, scientists often use a combination of these techniques to achieve separations that can be used for proteomics studies. Routinely, proteomists combine these techniques in a stepwise fashion. Also liquid chromatography (of peptides) has been coupled directly to tandem MS in a further attempt to achieve additional resolution.

7.2.4 Quantitative Proteomics

Although, mass spectrometry has been a very successful tool for studying proteins in complex mixtures, these studies have been so far dominated by qualitative results (Fig. 7.2). To complement this, proteomics researchers have developed two approaches to attain quantitative proteomic information. In general, a rough approximation of relative protein amounts between two samples can be extracted by comparison of the same peptide signals derived from samples prepared under different conditions. Alternatively, a predictable mass difference can be artificially introduced to more accurately accomplish quantitation. In order to prevent spectral overlap, incorporation of stable isotope labels should result in at least a 3 Da mass shift. These labels can be added using chemical "post-biosynthetic" or metabolic "pre-biosynthetic" approaches. The addition of the label allows for mixing of samples originating under different conditions for simultaneous analysis. When samples are mixed early in the workflow, less bias is introduced during sample processing, resulting in high reproducibility (Fig. 7.3). Therefore, methods that incorporate the stable isotope label at the protein level have higher reproducibility than those that introduce it at the peptide level. Different isotopic labeling techniques will be discussed in the following section.

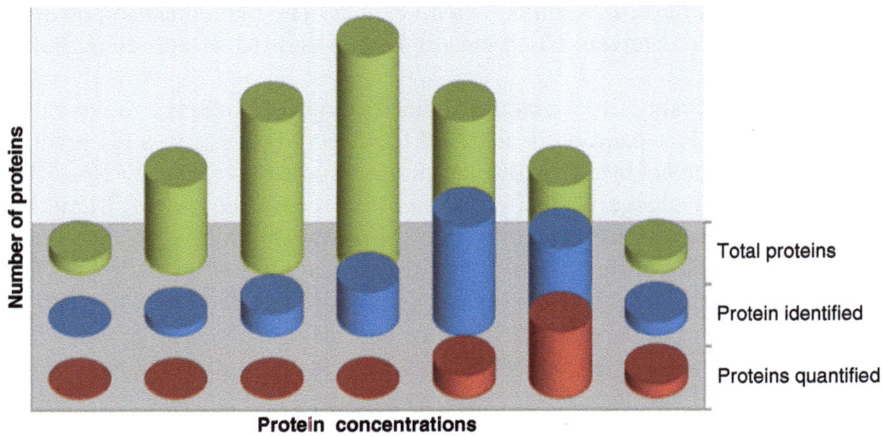

Fig. 7.2 Schematic depiction of identification and quantification of a proteome by mass spectrometry. Because of the limitations of current technology, only a fraction of the proteome can be identified, whereas a subset of that can be quantified

7.2.4.1 Pre-biosynthetic Labeling

In vivo labeling takes advantage of cell metabolism to effectively incorporate a stable isotope into proteins via the process of translation during cell growth and division. There are two approaches, the less practical is global labeling of proteins by growing cells by in ^{15}N-supplemented cell culture medium (Gygi et al. 1999). Although in vivo ^{15}N metabolic protein labeling of *C. elegans*, *Drosophila melanogaster* (Krijgsveld et al. 2003), rat (Wu et al. 2004), and plants (Gruhler et al. 2005) is feasible, it is not widely applied because it is time consuming and very expensive.

The most popular approach is the stable isotope labeling with amino acids in cell culture (SILAC) (Ong et al. 2002). In general, the stable isotope is incorporated by supplementing the cell growth medium with $^{13}C_6$-arginine and $^{13}C_6$-lysine. A second set of cells are grown in a label free environment. This approach guarantees that the resultant peptides from the tryptic cleavage of a protein do not overlap in the MS spectrum, excluding its C-terminus, since they contain no less than one labeled amino acid per peptide (heavy) with a constant mass increase of 6 Da as compared to the non-labeled corresponding peptides (light). The two cell populations are pooled, lysed, and proteins are isolated, denatured, reduced, and digested. The peptides are then quantified by MS. Protein identification is determined from either the "heavy" or the "light" peptide by MS, while relative quantitation is achieved by taking the ratios of the intensities of the two isotopes of the specific peptide in the MS spectrum. The advantage of SILAC over full metabolic protein labeling by ^{15}N lies in more straightforward data analysis since the labels in SILAC are specifically incorporated, defined, and not peptide-sequence dependent. SILAC is widely used for metabolic labeling of higher eukaryotic cells. Near complete incorporation of labels occurs after six doubling of cells grown in SILAC media (Ong et al. 2002).

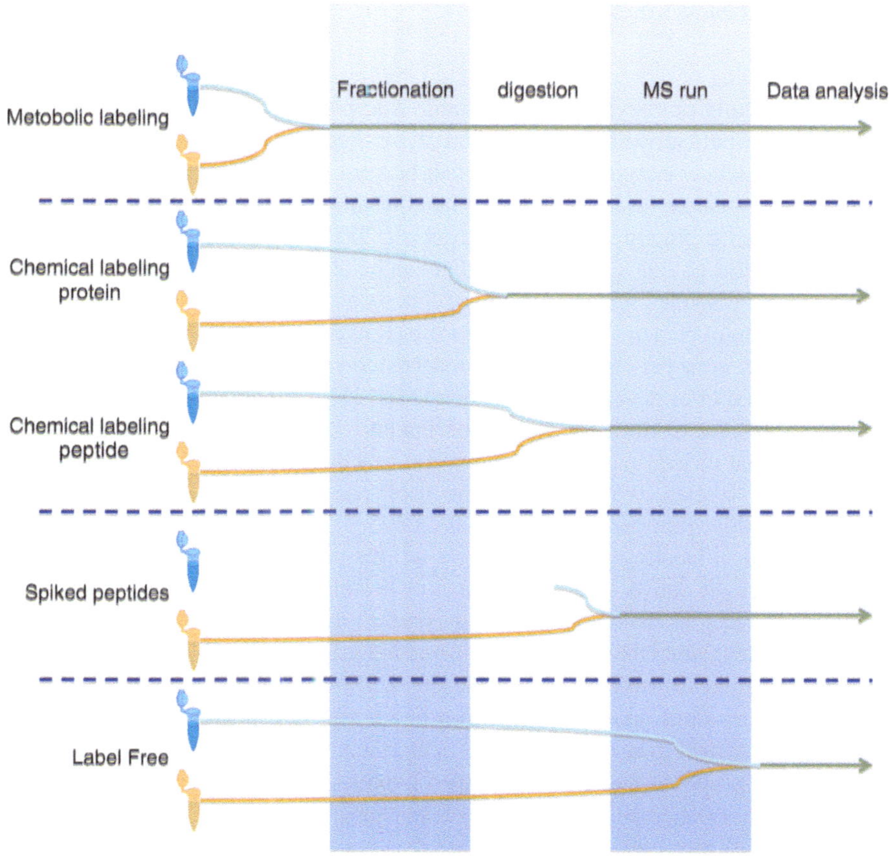

Fig. 7.3 Schematic diagram of the quantitative proteomics workflow. Tubes in blue and orange depict the two experimental conditions. The intersections of curved horizontal lines represent when samples are combined. *Shaded areas* represent different points of the experimental procedure and also where sample bias can be introduced if samples have not yet been combined

Although many cell lines can be labeled easily using SILAC, others are not. For instance, certain cell lines readily form proline from excess arginine, which can be alleviated by supplementing limited amount of arginine to the medium (Chelius et al. 2003). Some cell lines do not grow well in SILAC media and therefore cannot be labeled with this technique. Another limitation to the SILAC technology is the limited availability of useful isotopically labeled amino acids. As a consequence, in a single experiment only up to three conditions can be compared. For instance, the unlabeled sample can be compared to samples with $^{13}C_6$, $^{13}C_6$, and $^{15}N_4$ labels. One of the main advantages of SILAC specifically and metabolic labeling in general its reliable accuracy in quantitative MS-based methods due to early labeling and sample mixing. As a result, metabolic labeling is extremely useful for measurement of small variations in protein levels as well as post-translational modifications (Blagoev et al. 2004; Olsen et al. 2006; Park et al. 2006).

7.2.4.2 Post-biosynthetic Labeling

Isotopic labeling of extracted proteins and peptides can also be carried out in vitro either chemically or enzymatically. A stable isotope label can be incorporated into peptides enzymatically either during proteolytic digestion or in a separate step after proteolysis. Hence, enzymatic labeling can be very specific. For example, two ^{18}O isotope labels can be incorporated into the C-termini of peptides by either trypsin- or Glu-C during protein digestion (Yao et al. 2001; Reynolds et al. 2002). This results in a 4 Da (2 Da/^{18}O) mass shift that can be utilized for isotopomer discrimination. Other enzymes such as Lys-N introduce only one ^{18}O isotope and this mass difference cannot be detected in the spectrum (Rao et al. 2005). Since isotope labels can be lost at high pH (Schnolzer et al. 1996), electrospray ionization (ESI) and matrix-assisted laser desorption/ionization (MALDI) MS, which operate at moderate pH values, are utilized for these experiments. Another drawback of enzymatic labeling is that incorporation of isotopes is rarely complete and peptides are often differentially labeled, which can lead to tricky data interpretation (Johnson and Muddiman 2004; Ramos-Fernandez et al. 2007).

The chemical labeling approach mainly utilizes stable isotope-carrying chemical reagents to target active sites on peptides or proteins. The main targets for these reagents are the side chains of lysine and cysteine; therefore, this approach is not as specific as enzymatic labeling. The isotope-coded affinity tag (ICAT) was developed by Aebersold and co-workers (Gygi et al. 1999). The ICAT reagent targets and modifies cysteine residues and links them to a biotin tag by a polyether region, which contains either eight (heavy) or no (light) deuteriums. The biotin tag is used for affinity purification and recovery of the labeled peptides. The ICAT experiment is performed on two isolated populations of proteins that are reduced and tagged with light and heavy ICAT reagents. The proteins are then pooled and digested and the tagged peptides are recovered by affinity chromatography and quantified by MS.

ICAT-generated samples are less complex than those obtained from other chemical approaches since only cysteine, a rare amino acid, is labeled and thus analysis of complex samples is feasible. However, this also means that proteins with one or no cysteines are not detectable. In addition, the large tag effect on the fragmentation spectra and eluting time during reverse phase chromatography between lights and heavies are troublesome. These limitations have been overcome by recent technology advances, such as replacing the linker with one that is cleavable (Hansen et al. 2003; Li et al. 2003; Oda et al. 2003). A similar method makes use of a 2-thiopyridyl disulfide group to react with cysteines, a deuterium-labeled alanine, a His_6-tag for affinity purification, and a tryptic cleavage site to limit the size of the tag (Olsen et al. 2004). ICAT and similar methods are valuable tools for a host of expansive, human plasma, or targeted analyses.

Chemical labeling can also be achieved by a group of reagents that modify the N-terminus of the peptide as well as the epsilon-amino group of lysine residues. The most common and specific reagents are the *N*-hydroxysuccinimide (NHS) and other active esters and acid anhydrides. This group includes isotope tags for relative and absolute quantification (iTRAQ) (Ross et al. 2004), the isotope-coded protein label (ICPL) (Schmidt et al. 2005), tandem mass tags (TMT) (Thompson et al. 2003), and acetic/succinic anhydride (Glocker et al. 1994; Ji et al. 2000; Che and Fricker 2002; Zhang et al. 2002). Less commonly employed reagents are isocyanates, isothiocyanates (Mason and Liebler 2003; Lee et al. 2004), and formaldehyde methylation of lysine residues followed by reduction by cyanoborohydride (Hsu et al. 2003; Ji et al. 2005; Hsu et al. 2006).

In these chemical labeling methods, labeled and unlabeled peptides are mixed and then quantification is performed by using the ratio of the MS signal intensities of the different isotopes or reporter ions. One advantage of the abovementioned methods is the use of isobaric tagging of peptides (Thompson et al. 2003). This results in peptides that co-elute during liquid chromatography, leading to reduced variability. The different tags are then distinguished by the mass spectrometer after fragmentation occurs. For instance, in single MS mode the same peptides with different labels are identical in mass. However, in tandem MS mode, where the peptides are fragmented, each tag generates a unique reporter ion. Protein quantitation is then achieved by taking the ratio the intensities of the reporter ions relative to each others in the MS spectra. This approach allows the simultaneous determination of both identity and relative abundance of peptide pairs in MS. The main advantage of the iTRAQ reagent (Ross et al. 2004) is that it allows multiple quantitation of up to eight samples at the same time thereby reducing the amount of mass spectrometry time needed for analysis.

Another type of chemical isotopic labeling targets the carboxylic acids in proteins. The C-termini of proteins as well as glutamate and aspartate are esterified by deuterated alcohols (David et al. 2001; Syka et al. 2004b). This approach has proven useful in quantitative studies of phosphorylated peptides, since it reduces the cross-reaction with ion metal chelate affinity chromatography (IMAC) mentioned earlier (Salomon et al. 2003). Also β-elimination of phosphoric acid followed by Michael addition is used for quantitation studies of phosphorylated peptides (Goshe et al. 2001, 2002; Qian et al. 2003; Tao et al. 2005). Quantitative studies of glycosylated peptides are achieved by the usage of hydrazide-based reactions, in which the carbohydrate is replaced by an isotopically labeled tag (Zhang et al. 2003).

All chemical labeling methods can be applied to proteins as well peptides. For example, labeling of the N-termini and lysine side chains of proteins has been applied using iTRAQ and ICPL. An advantage of labeling at the protein level is the minimization of bias introduced at later steps, since sample combination can be achieved early in the process. However, there is a drawback to protein labeling: Trypsin cannot cleave modified lysine residues, resulting in lower identification coverage due to presence of long peptides.

7.2.4.3 The Absolute Quantification Strategy

The absolute quantification of proteins (AQUA) strategy (Gerber et al. 2003) is accomplished by adding modified peptides to a sample as internal standards.

These peptides can contain stable isotopes and can be synthesized with covalent attachments to mimic protein posttranslational modifications such as phosphorylation, methylation, and acetylation. Data analysis is performed by comparing the signal from the synthetic peptide to the native peptide in the MS spectrum. The AQUA approach is limited to the quantitation of only a small subset of any sample, but it is still very useful if the aim of the study is focused on one or few proteins. For example, Gerber et al. (2003) measured the cell cycle-dependent modification of the human separase protein. To alleviate the limitations of AQUA, a *de novo* gene design was developed to express artificial proteins that are concatemers of tryptic Q peptides (QCAT) (Beynon et al. 2005). This strategy increases coverage, reduces bias, and provides better accuracy due to the introduction of the peptides early in the process. This strategy was applied successfully in the absolute quantification of the components of the eIF2B–eIF2 protein complex (Kito et al. 2007).

A limitation to the AQUA and similar strategies is underscored by the inherently narrow dynamic detection range of present mass spectrometry, which is compounded by the complexity of the tryptic digests of entire proteomes. The amount of labeled standard that must be spiked into the sample is rather difficult to determine since proteins of interest can be expressed differentially under diverse conditions. Also the specificity of the added standards is potentially problematic if they result in multiple isobaric peptides. The use of a very clever strategy called multiple reaction monitoring (MRM) can alleviate both of these limitations (Kirkpatrick et al. 2005). In an MRM experiment, a triple quadrupole mass spectrometer is employed to facilitate two stages of mass filtering. The intact ionized peptide is preselected and fragmented and then a small number of resultant sequence-specific fragment ions are mass analyzed. This targeted MS analysis using MRM improves specificity in peptide assignments, expands the quantitation scale, and enhances the detection limit for peptides by up to 100-fold (Wolf-Yadlin et al. 2007). Also the choice of tryptic peptides to be used can be assisted by use of a platform that predicts the most likely protein fragments to be observed, hence facilitating the choice of standard peptides (Mallick et al. 2007).

7.2.4.4 Label-Free Quantification

Proteomic quantification can also be achieved without artificially labeling parts of the sample. There exist two approaches to accomplish this label-free quantification: extraction of peptide ion intensities (Bondarenko et al. 2002; Chelius and Bondarenko 2002; Chelius et al. 2003; Wang et al. 2003; Li et al. 2005) and spectral counting (Gao et al. 2003; Liu et al. 2004). The first approach is based on comparing integrated areas under the curve of extracted peptide ion intensities (Higgs et al. 2005). The accuracy of this method is limited by the mass accuracy and reproducibility of the mass spectrometer. To achieve high accuracy, one should minimize the signal overlap by utilizing a high mass accuracy spectrometer. Also, utilization of LC alignment software can optimize the chromatographic profile of peptides (Bylund et al. 2002; Strittmatter et al. 2003; Jaitly et al. 2006; Wang et al. 2007) in turn enhancing reproducibility. These types of experiments require an immense amount of time; therefore, a compromise has to be made between identification and quantitation. As a consequence, better quantification accuracy is achieved at the expense of coverage and vice versa.

An alternative is the spectral counting approach, which depends on wide data acquisition for both identification and quantitation. The spectral count approach is relatively new and relates the number of mass spectra identified for a protein to the protein's abundance (Gao et al. 2003; Liu et al. 2004; Gilchrist et al. 2006). Therefore, a direct comparison of two or more runs will allow the relative quantification of the protein of interest. To achieve better accuracy and reliable quantitation, an exponentially modified protein abundance index (emPAI) (Ishihama et al. 2005) is utilized, which is proportional to concentration of proteins in a sample. In addition, better quantitation is achieved through use of computational tools that select peptides in advance for detection by the mass spectrometer (Craig et al. 2005; Tang et al. 2006; Lu et al. 2007; Mallick et al. 2007). The minimum number of spectral counts required to see a significant change was determined by Old et al. (2005); they observed that the relationship is not linear, but rather exponential. They concluded that four spectra were sufficient to see threefold protein changes, but up to fifteen spectra were needed to observe a twofold change. They also showed that the spectral counting method yields reliable results as compared with extraction of peptide ion intensities, but both methods are less sensitive than isotopic labeling (Old et al. 2005). Although this approach has a great benefit highlighted by the simultaneous protein identification and quantitation, it suffers a major drawback. Quantitation is greatly dependent on the quality of MS/MS peptide identification, since errors in peptide identification can lead to inaccurate protein quantitation (Li et al. 2003; Olsen et al. 2006). Although both label-free methods have their advantages and can be applied for global quantitation studies, both require extensive platform setup. Hence only a handful of labs are able to take advantage of these methods.

7.3 Mass Spectrometric Analysis

As noted above, technological advances over the last 20 years have made mass spectrometry the tool of choice for proteomics researchers. In the late 1980s mass spectrometry of biological samples was improved by two novel ionization techniques, MALDI and ESI (Fig. 7.4). After that, the field further developed with advances in sample preparation, instrumentation, and sample analysis algorithms; all of which will be discussed in this section. In general, a mass spectrometer is used to answer two questions about a biological sample: "What is in the sample?" and "How much is in the sample?" Our goal with the following section is to provide insight into how mass spectrometers function so as to improve understanding of the resulting values of identity and quantity that are critical to modeling biological systems. As noted above, the most common proteomics paradigm uses mass

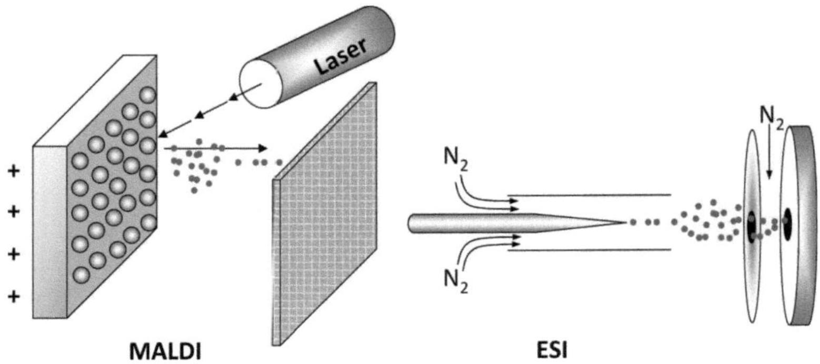

Fig. 7.4 Schematic of ionization methods. In *MALDI*, a sample is co-crystalized in a matrix solution atop a target plate. The prepared sample may then be irradiated with a laser, resulting in a vapor phase, then accelerated away from the source due to a high potential applied to the source. In *ESI*, a sample dissolved in solution passes through a highly charged needle and then passes to the inlet of the mass spectrometer, sometimes with the aid of nitrogen gas

spectrometry on polypeptide fragments of proteins as produced by digestion of a protein mixture with trypsin.

Fundamentally, mass spectrometers measure the molecular mass of a polypeptide and additional structural information, such as amino acid sequence or post-translational modifications, can be inferred. In its most basic structure, a mass spectrometer has three functions: (1) ionization, the production of gas-phase ions from the sample; (2) mass analysis, the separation of gas-phase ions according to their mass-to-charge (*m*/*z*) ratio; and (3) detection of separated ions. Initially the production of gas-phase ions proved difficult for biological samples due to "excessive" fragmentation until the advent of MALDI and ESI. These two "soft ionization" methods made it possible to generate ions from intact biomolecules.

7.3.1 Ionization

7.3.1.1 MALDI

MALDI takes advantage of lasers and matrix material to generate charged ions. Protein samples are dissolved in a matrix solution (typical compounds used are α-cyano-4-hydroxycinnamic acid or dihydrobenzoic acid). Samples in submicroliter to microliter volumes are allowed to dry on a metal substrate. After drying, samples are irradiated with nanosecond laser pulses. The matrices that dissolve the samples differ in energy necessary to desorb from surface. The matrix also affects the amount of fragmentation the samples undergo. The energy necessary to desorb is inversely correlated with ion stability. Methods have been developed to increase the stability of peptides in the ionization process, for example, the addition of nitrocellulose increases the representation of peptides (Jensen et al. 1997). α-Cyano-4-hydroxycinnamic acid matrix solutions generally lead to highest sensitivity in MALDI for biological samples.

7.3.1.2 ESI

In ESI, samples remain in liquid form and the analytes are pumped at submicroliter to microliter per minute flow rates through a needle that is under high voltage. This voltage electrostatically disperses the sample into droplets that evaporate into charged vapor droplets. A sheath gas is used to aid in the transfer of the vapor into the mass spectrometer. This technique has the advantage of being gentler on polypeptides than MALDI. It can also be used in tandem with liquid chromatography. To be analyzed by ESI-MS, molecules must have sufficient polarity to allow attachment of a charge. The signal strength, which is essentially the peak height in the spectrum, increases linearly with the analyte concentration over a wide range until saturation occurs. There does not seem to be an upper mass limit to analysis by ESI-MS. Large ions, like proteins, are typically and are therefore in the range of mass-to-charge (*m/z*) ratios of typical mass spectrometers. The distribution of charges gives rise to a multiple charge envelope but spectra can be simplified by deconvolution, an algorithm that sums up the signal intensity into a single peak at the molecular weight of the analyte. Very complex mixtures can be analyzed by ESI-MS, but the spectra become increasingly difficult to interpret with increasing molecular weight and numbers of compounds.

Each ionization technique has advantages. ESI may be coupled to liquid chromatography systems. The investigator may use various gradients and separation techniques (e.g., C18 columns) to separate peptides prior to ionization to enhance the resolving power of the mass spectrometer. However, unlike the MALDI ionization set-up, the investigator may not go back to source (re-gain the sample) in order to look at it again. For ESI, the entire experiment would have to be re-run with a comparable sample. In MALDI, the samples are usually added to the matrix in spots on the target and the laser usually only irradiates a small area of that spot. Therefore, the investigator can analyze the same sample again and again. This has the advantage of increasing the mass accuracy and confidence in the data.

Electrospray ionization is the most common form of ionization coupled to Fourier transform mass spectroscopy (FTMS) and has the advantage of being configured with a nanospray source. Typical specifications for a nanospray set-up include flow rates on the order of 250 nL/min and sensitivity in the femtomolar range. Using an LC coupled to the mass spectrometer, the investigator can run a number of experiments in-line with the instrument. Complex mixtures (e.g., serum) can be separated on a column based on protein chemistry prior to ESI-MS analysis. Additionally, in some cases, samples may be fractionated into *n* numbers of fractions prior to MS analysis using the techniques described above. Each fraction can then further be separated on a column in-line with the instrument resulting in lower complexity fractions and therefore more proteins/peptides identified.

7.3.2 *Mass Analyzers*

The function of the mass analyzer is to separate ions based on their mass-to-charge ratio. A number of types are available and will be discussed in this section (Fig. 7.5).

7.3.2.1 Quadrupole Mass Analyzer

The most commonly used mass analyzer is the quadrupole mass analyzer which is often referred to as a mass filter. This analyzer has four adjacent metallic rods that

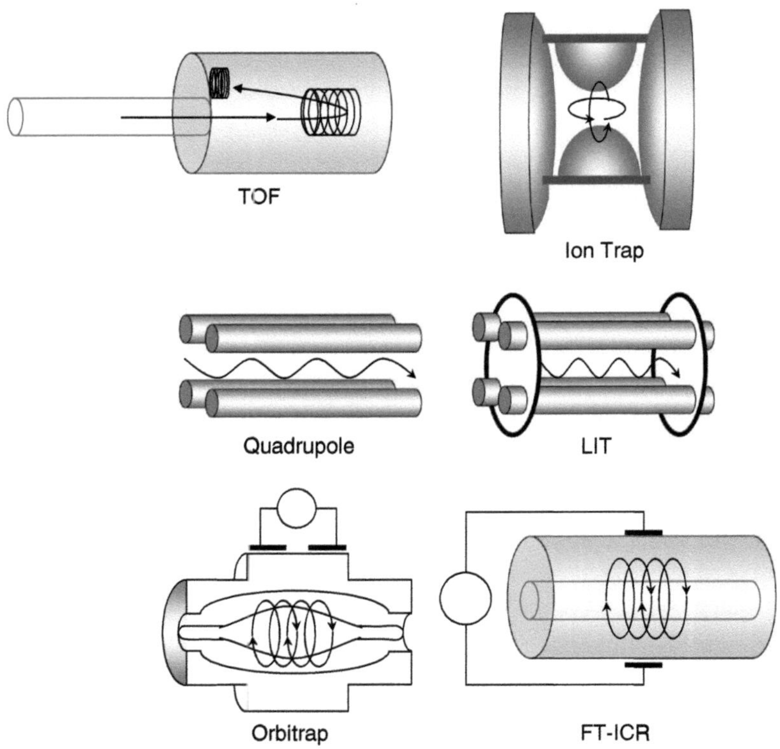

Fig. 7.5 Schematic of common mass analyzers. The *TOF* mass analyzer has a reflectron at the end to correct for shifts in flight times. The *Ion Trap* is used to perform MS/MS, can be used in tandem, and is often connected to ESI. Of the instruments available it has a low mass accuracy and resolution. *Quadrupole* mass filters can operate in tandem; in the triple quadrupole, the ion activation often occurs in the second quadrupole. *LIT* is similar to the quadrupole; it has endcaps (with DC potentials) to allow ion trapping along the long axis. The *Orbitrap* is another relative of the ion trap except with resolution and mass accuracy comparable to FT-ICR. *FT-ICR* mass spectrometers provide the highest resolution and mass accuracy of mass spectrometers available

are connected pair wise; each pair is set to a positive or a negative electrical potential. A combination of direct current (DC) and radio frequency (rf) voltages are applied between the rods in order to move the ions through the quadrupole. Depending on voltage, only ions of a given *m/z* value travel along the analyzer to the detector, while other ions collide with rods and are lost. By scanning the DC and RF voltages, while keeping the ratio constant, ions with different *m/z* ratios pass through to the analyzer successively so that a wide *m/z* range may be scanned (March 1997; Schuchardt and Sickmann 2007). With this design, ions may be "trapped" in a defined volume (i.e., ion trap) or drift downstream into other cells, such as other quadrupoles (tandem mass spectrometry will be discussed further below).

7.3.2.2 Time of Flight

The time of flight (TOF) mass spectrometer has a simple design: The ions are accelerated across a "field-free-drift region" of the flight tube and the velocity of the ions in the analyzer tube is dependent on their *m/z* values. The typical set-up has the ions traveling through a flight tube and then reflected at the end to a detector by an ion mirror called a reflectron (Karas and Hillenkamp 1988). This set-up is preferred because the alternative, the linear tube design, has relatively poor mass resolution. The reflector at the end of the flight tube is used to correct for initial energy differences; it corrects for the error in flight times by focusing the ions with the same *m/z* in space and time before they hit the detector. With the reflectron TOF, resolution up to 25,000 is easily achieved. MALDI ionization is often combined with time-of-flight (TOF) analyzers.

7.3.2.3 Ion Trap

In ion trap analyzers the ions are trapped in a cell for a certain time interval and then subjected to MS or MS/MS analysis. These ions are trapped using electric fields and limit for the amount of ions trapped is based on their space charge (Louris et al. 1987). The maximum number of ions is just below the number that distorts the applied field. The ions are then subjected to another electric field that ejects ions from the trap, resulting in a mass spectrum. For MS/MS, the unwanted ions are ejected first then the ions of interest are fragmented further and analyzed. Ion trap mass analyzers provide fast scanning rates, sensitivity, flexibility, and robustness and have the advantage of relatively low cost.

Ion traps have two primary designs. One is the quadrupole ion trap (QIT) and the second is the linear, or 2D, ion trap (LIT). The QIT is structurally a quadrupole analyzer that uses a combination of the rf and DC voltages to select ions of a particular *m/z*, but ions are trapped in the three dimensions of the cell. By ramping up the rf voltages, the QIT moves the ions out of the trap to the detector and the spectrum is scanned. These ion traps are ideal for MS/MS experiments, since the trapped ions may be excited via collision-induced dissociation (CID) to generate the fragment (or MS/MS) spectrum. The linear ion trap is similar to the QIT, except that there are additional DC potentials to allow ions to be trapped along the long axis of the quadrupole. Ions are ejected either radially (as in the Thermo-LTQ) or axially (as in the ABI/Sciex Q-Trap) through a series of ramping protocols (Khalsa-Moyers and McDonald 2006). The primary advantage of QIT over LIT is that QIT has a greater trapping volume and therefore analyzes more ions per cycle. This improves the sensitivity and dynamic range of the ion trap.

7.3.2.4 Ion Cyclotron Resonance

One of the most powerful mass spectrometers on the market is the Fourier transform ion cyclotron resonance (ICR) mass spectrometer (FT-ICR or FTMS). The FT-ICR is a trapping mass spectrometer that captures ions under high vacuum in a high magnetic field. ICR was developed by Comisarow and Marshall in 1974 (Comisarow and Marshall 1974). In the ICR, ions travel forward and rotationally through the mass spectrometer (similar to a corkscrew motion) under high vacuum and within a magnetic field. The applied field resonates with the ions and as their "cyclotron" path is widened the rotational speed is measured in order to determine the ion size. The FT-ICR mass spectrometer was a breakthrough in resolving power and mass accuracy (Senko et al. 1997; Domon and Aebersold 2006). The FT-ICR provides high quality data and allows the detection of more signals than do instruments of lower resolving power. The development of a hybrid FT-ICR instrument with an external LIT device allows parallel full mass spectrum (MS1) and tandem mass spectrum (MS2) acquisition (not sequential); the high-quality MS1 data can be used for quantification. The system is limited by a relatively slow acquisition rate (several s per cycle).

7.3.2.5 Orbitrap

Orbitraps are similar to ICR mass spectrometers, except that rather than using a magnetic field, an electric field is used (Makarov 2000; Hardman and Makarov 2003). The Orbitrap radially traps ions about a central spindle electrode. The outer barrel-like electrode is coaxial with the inner spindle-like electrode and mass/charge values are measured from the frequency of harmonic ion oscillations of the orbitally trapped ions. Ion frequencies are measured non-destructively by acquisition of time-domain image current transients (Makarov 2000; Schuchardt and Sickmann 2007). In simpler terms, instead of measuring the ions' rotational frequencies, the translational motion along the long axis of the Orbitrap cell is measured. The Orbitrap provides high resolution (up to 150,000) and high mass accuracy (2–5 ppm) (Hardman and Makarov 2003; Hu et al. 2005). Even though, the FT-ICR provides better resolution and mass accuracy, the Orbitrap is an attractive alternative for most users and applications because there is no need for liquid helium or liquid nitrogen, as there is with the large superconducting magnet in the ICR. Similar to other mass analyzers, it may be combined with either MALDI or ESI sources.

7.3.3 *Using a Mass Spectrometer to Identify Species in a Mixture*

Given that tandem mass spectrometry is so powerful, how exactly is it used to identify ions and assign them to peptides? How is sequencing done in a mass spectrometer? Here we describe the process of sequencing using the mass spectrometer (Fig. 7.6).

7.3.3.1 Collision-Induced Dissociation

In the CID experiment, precursor ions are isolated and subjected to a neutral target gas. The gas collides with the precursor ions passing-on kinetic energy to the ions (Sleno and Volmer 2004). Multiple collisions increase the internal energy of the ions, resulting in fragmentation of the peptide backbone primarily at the amide bonds (Sleno and Volmer 2004). This results in b and y ions (Fig. 7.7). In CID, both low and high energies are used to fragment the ions. Low-energy collisions are used in ion trap and quadrupole instruments. This low-energy fragmentation results in a, b, y, and immonium ions, and ions from the neutral loss of ammonia. In the high-energy CID, the ions observed are d, v, w, and immonium ions. For mass spectrometers such as triple quadrupoles, ion traps, or TOF instruments, ions are isolated and

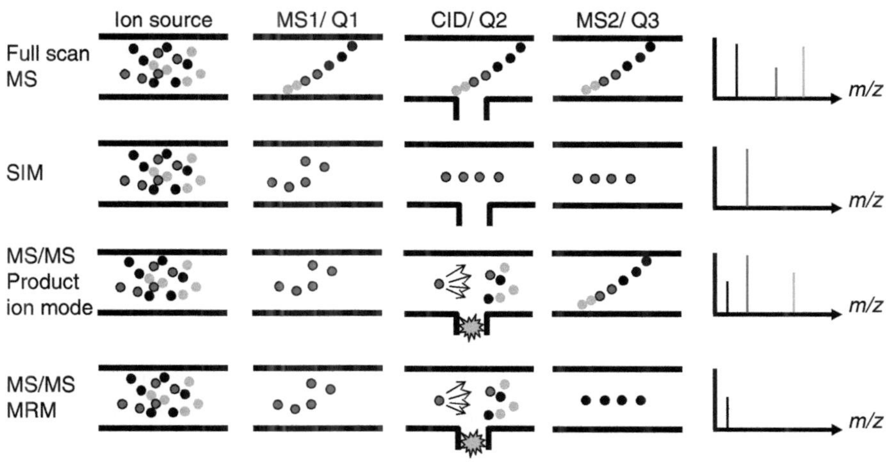

Fig. 7.6 Schematic of the different types of tandem mass spectrometry experiments, including the triple quadrupole experiments. Both *Full Scan MS* mode and *SIM* are examples of experiments using single quadrapole mode. Full Scan mode is a simple scan of all the ions from the ion source. In *SIM* mode, an ion is selected in the first quadrupole and then scanned. In *Product Ion Mode*, a precursor ion is selected in Q1, then activated in Q2 (collision induced dissociation region), and then scanned in Q3. The *MRM* mode may be a series of experiments that first select precursor ions and then filter the resulting fragment ions post Q2. The experiment may be set up to look for multiple resulting fragment ions in Q3

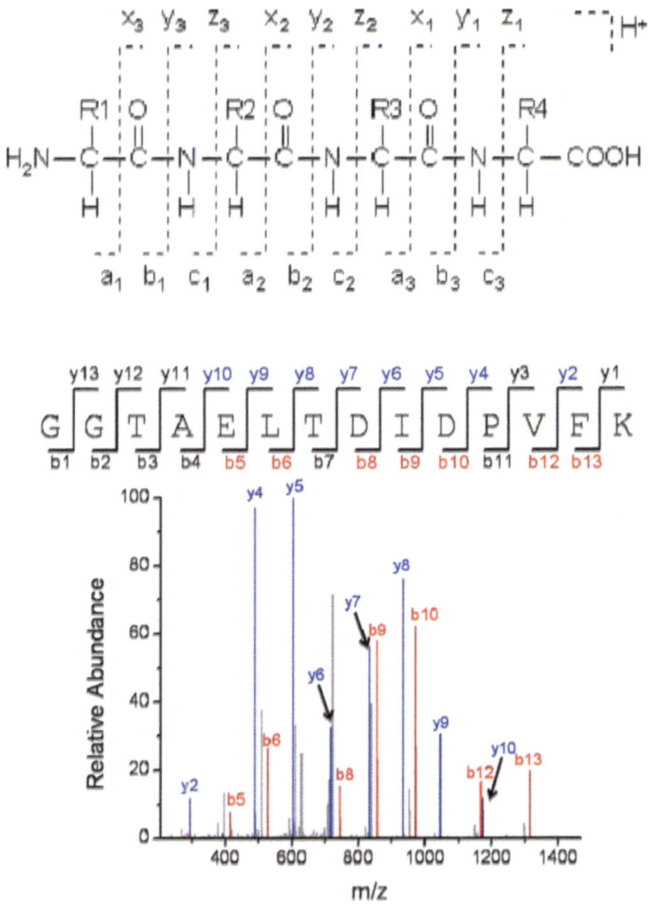

Fig. 7.7 Peptide fragment ion nomenclature. Resultant ions from cleavage of bonds along the peptide backbone. Typically b- and y-ions result from CID whereas c- and z-ions result from ECD

fragmented in a collision cell. In tandem mass spectrometers, such as QQTOF, ions enter the first quadrupole, ions of interest are then sent to the second quadrupole and the cell is filled with target gas. The resultant fragments are then sent downstream to a TOF detector.

7.3.3.2 Electron Capture Dissociation

Electron capture dissociation (ECD) was first introduced by Zubarev et al. (Zubarev et al. 1998, 2000). For ECD, a low-energy electron beam (<0.2 eV) generated by a heated filament electron gun is used for activation (Zubarev et al. 1998; Cooper et al. 2002). The positively charged precursor ions capture the electron leading to neutralization and backbone fragmentation. Backbone cleavage usually occurs at the N–Cα bond yielding primarily c and z ions (Zubarev et al. 1998). ECD preferentially cleaves disulfide bonds but leaves other post-translational modifications intact (Zubarev 2004). ECD is an available option on most FT-ICR instruments and has been used to characterize post-translational modifications such as O-linked glycosylation, methionine oxidation, and phosphorylation (Bakhtiar and Guan 2005; Zhang et al. 2005b).

7.3.3.3 Electron Transfer Dissociation

The ECD activation method is not amenable to use in ion trap instruments. However, an analogous ion-ion method called electron transfer dissociation (ETD), where ion–ion reactions occur between singly charged anions and multiply charged peptide cations, was developed by Hunts and colleagues (Syka et al. 2004a). The electron source in ETD is a chemical ionization. The anions are introduced into the trap via an anion beam controlled by RF gating voltages. The anions interact with the multiply charged peptides resulting in a proton transfer without dissociation and in electron transfer with or without dissociation. Proton transfer results in charge reduction, whereas dissociation leads to c- and z-fragmentation, quite similar to what is observed in ECD (McLafferty et al. 2001).

7.3.3.4 Infrared Multiphoton Dissociation

Infrared multiphoton dissociation (IRMPD) is a slow heating dissociation method involving non-resonant ion activation and subsequent dissociation via photon absorption (Khalsa-Moyers and McDonald 2006). This method was historically used for small molecule analysis, but IRMPD has recently been applied to protein analysis (Shukla and Futrell 2000; Sleno and Volmer 2004). In the IRMPD experiment a low-powered CO_2 laser is used to activate ions. This laser is useful for analyzing phospho-peptides, because the phosphate preferentially absorbs at this wavelength. Since IRMPD is a low-energy ionization, it shows similar fragmentation patterns to CID (Zhang et al. 2005b).

7.4 Data Analysis

7.4.1 Identification

Peptide sequencing and identifying the peptides being fragmented in the mass spectrum is key to mass spectrometry and proteomics. How these fragmented ions are identified can be organized into three main categories: (a) database searching, similar to spectral library searching, where peptide sequences are identified based on theoretical spectra predicted for that sequence or based on spectra from previous experiments; (b) *de novo* sequencing, where peptide sequences are read out directly from fragment ion spectra; and (c) hybrid techniques, where short stretches of the peptides are sequenced then the rest of the spectrum is searched through databases. For a comprehensive review of the publicly available tools for MS/MS-based proteomics, see (Nesvizhskii et al. 2007).

Peptide mass fingerprinting (PMF) is considered one of the fastest methods for identifying proteins recovered after gel electrophoresis or other isolation methods that provide samples containing one or two proteins. In PMF, the protein of interest is isolated from a gel and digested with a proteolytic enzyme (e.g., trypsin which selectively cleaves the protein at lysines and arginines) (James et al. 1993; Mann et al. 1993; Pappin et al. 1993). The mass spectrum obtained from a MALDI-TOFF is then searched against the masses from a known protein/peptide databases. This method has the advantage of speed, it is much faster than the labor intensive *de novo* sequencing, however it is only effective if the protein in question has actually been sequenced and is in the database.

In database searching, the spectrum of a protein is scored against theoretical fragmentation patterns constructed for peptides found in the searched databases. The peptides queried are restricted to investigator specified criteria (e.g., proteolytic enzyme and post-translational modifications allowed). Once a spectrum is matched against spectra from the database, a list of ranked peptides (scored according to the parameters set by investigator) is returned. Discerning a true match from a false match is critical in proteomic data analysis. The higher the score the more confident the investigator is that the peptide is a positive match. There are a number of scoring schemes: spectral correlation functions (e.g., SEQUEST) (Eng et al. 1994), shared fragment counts and dot products (e.g., TANDEM, OMSSA, MASCOTT) (Perkins et al. 1999; Craig and Beavis 2004; Geer et al. 2004), empirically observed rules (e.g., Spectrum Mill), and fragmentation frequencies (e.g., PHENYX) (Colinge et al. 2003). These scores can be converted into an expectation value (E value), which is the expected number of peptides with scores equal to or better than observed score under the assumption that peptides are matching the experimental spectrum by random chance (e.g., OMSAA, TANDEM and MASCOT).

Despite success of database and spectral matching searching, false peptide assignments occur for a number of reasons. Reasons for false assignments are use of simplified scoring algorithms, contaminants, low quality spectra, fragmentation of multiple peptide ions, presence of homologous peptides, incorrectly determined charge state or peptide mass, restricted/limited database search, sequence variants and new peptides (Nesvizhskii et al. 2007). The generation of "high confidence" identifications is the goal in proteomics, but scoring is software/tool-dependent. The score distribution depends on mass spectrometer performance, the quality of the sample, the instrument settings and its methods, and the size of the database. The quality of the score may be improved by approaches such as target-decoy searching and use of empirical Bayes methods (Keller et al. 2002; Storey and Tibshirani 2003; Elias and Gygi 2007). The target-decoy method is when the peptide in question is searched against the database of peptides in reverse order or with "shuffled" sequences. Subsequently, peptides are filtered with score cut-offs. This method is useful in "weeding" out false positives but has the disadvantage of requiring twice the computing time. Programs such as PeptideProphet employ empirical Bayes approaches to validate peptide assignments made by database search programs. From each dataset, it learns distributions of search scores and peptide properties among correct and incorrect peptides and uses those distributions to compute probabilities that assignments are correct.

Identifying peptides based on their spectral match to a spectrum in a spectral library has been expanded because of the in-depth coverage of proteomes in eukaryotic species (Brunner et al. 2007; Brill et al. 2009). With extensive maps already in existence, the expansion to the proteomes of other systems by spectral matching to spectral libraries will grow at a much faster rate (Yates et al. 1998; Craig et al. 2006; Frewen et al. 2006; Lam et al. 2007).

In *de novo* sequencing, amino acids in the peptide are directly read from the fragment ion spectrum, facilitated by tools such as PepNovo and PEAKS (Ma et al. 2003; Frank and Pevzner 2005). Direct sequencing is helpful in cases where peptides are modified or there are polymorphisms. When limited genome information is known about a host, *de novo* sequencing or a hybrid of *de novo* with database spectral searching must be used. Posttranslational modifications (PTMs) are best identified through direct *de novo* sequencing; however, there are database searches and hybrid searches that can be employed to identify PTMs. A prominent inefficiency in the shotgun approach to proteomics lies in the redundancy of peptides seen. In complex samples as the human serum proteome, laboratory measures such as fractionation and immunological depletion experiments are done to compliment the analytical and software approach of the mass spectral data. Combining several search scores improves the overall confidence of the peptides identified. Programs such as TANDEM allow investigator-provided constraints, to allow for stricter and more confident identifications. Auxiliary run-condition information can also improve spectral identification. For example, the retention time (Strittmatter et al. 2004) and/or known sequence motifs such as the presence of N-linked glycosylation sites (Zhang et al. 2005a) can be useful.

7.4.2 Quantification

MS based methods for quantitative analyses of proteins/peptides, as discussed earlier, seek to compare two or more distinct proteomes in order to identify proteins with altered expression levels or post-translational forms in response to a given stimulus. In quantification analysis, regardless of whether the samples are isotopically labeled or label free, the ratio of intensities of the peptide peaks in a given mass spectrum gives a relative ratio of abundance of the two or more species. Several factors have to be considered when performing quantitative experiments. When choosing a stable isotope label, it must be determined whether the label alters the physicochemical properties of a peptide. For example, there is minimal impact when using ^{13}C, ^{15}N, or ^{18}O labeling (Zhang and Regnier 2002), but deuterium labeling can be problematic as labeled and unlabeled peptides differ in their retention time in RP-HPLC (Zhang et al. 2001). This results in an inaccurate quantitation data analysis and requires an additional signal integration step over retention time to correct for the inaccuracy (Fig. 7.8).

As mentioned earlier, to prevent spectral overlap, the stable isotope label incorporation should result in at least 4 Da shift relative to the unlabeled peptide. Another area of great effect on the accuracy of quantification is the quality of spectrum. Data should be handled with scrutiny when the signal is very low (close to the noise level) or very high (possibly resulting in detector saturation) as both will lead to distortion of the isotope envelope intensity leading to inaccurate quantitation. It is also dependent on the ability of the instrument to discriminate between interfering signals resulting from co-eluting peptides and the peptide isotope envelopes. Even though this can be minimized by reducing the sample complexity through fractionation, it should be noted that analytes often do not elute in a narrow profile and sometimes even elute into two or more fractions in separated regions of the elution profile (Faca et al. 2007).

In MS/MS based quantitation studies, the detector saturation problem is minimal and quantitation is not dependent on the machine's mass resolution, but on the size of the sequencing window, therefore it is background contributions that may bias the results. Hence, the *m/z* window used for sequencing should be optimized for every run. When employing the spectral counting technique, results can be computed in any of several ways. The simplest reports the average of ratios (Saito et al. 2007) while using an intensity threshold in order to minimize the noise based bias (Wolf-Yadlin et al. 2007). More reliable results are achieved when the ratios are computed based on the intensity weighted average, on the sum of all the observed spectra (Ono et al. 2006; Saito et al. 2007), or by employing linear regression (Parish 1989).

7.5 Applications

In the previous section we described the technology underlying proteomics approaches. Here we briefly describe two primary applications of proteomics to biology. First we discuss the identification of post-translational modifications. Next, we describe how proteomics approaches can be used to characterize protein complexes.

7.5.1 Post-translational Modifications

Post-translational modifications (PTMs) are covalent processing events that change the properties of a protein by proteolytic cleavage or by addition of a modifying group to one or more amino acids. PTMs can determine a protein's activity state, localization,

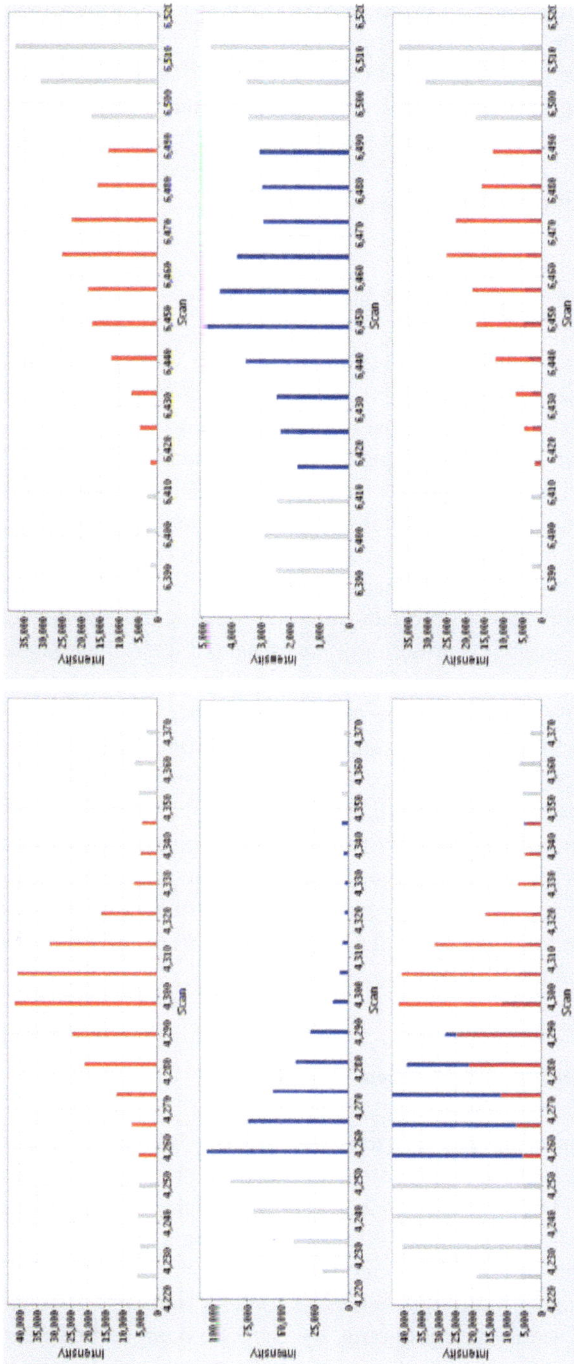

Fig. 7.8 Bias introduced by non co-eluting peptides. The elution profiles (*left*, non co-eluting; *right*, co-eluting) of peptides to be quantified and the potential of biasing the results. *Top* and *middle* represent extracted ion scans of one peptide and its isotopically labeled counterpart. *Bottom* two scans are the combination of the differentially labeled peptides

turnover, and interactions with other proteins. For example, kinase cascades critical to signaling are turned on and off by the reversible addition and removal of phosphate groups and cyclins are marked for destruction at defined time points in the cell cycle by ubiquitination. Despite the great importance of PTMs for biological function, their study on a large scale has been hampered by a lack of suitable methods and many key modifications have only been discovered late in the elucidation of various biological processes. As a result, we probably do not realize the full extent and functional importance of protein modifications in the workings of the cell.

7.5.1.1 Isolation of Modified Proteins

Modification analysis is usually done by comparison of experimental data to a known amino acid sequence. Therefore, the first step is identification of the protein to be studied, which can be done at very high sensitivity by antibody recognition (Western blotting) or by MS techniques. A central consideration in the characterization of modifications is the need for as large an amount of the protein as possible. Protein modifications are typically not homogeneous and a single gene may give rise to a bewildering number of gene products as a result of alternative splicing and the combination of different post-translational modifications. The amount of protein in a single modification state can thus be a very small fraction of the total amount of the gene product. Furthermore, as explained later, the complete characterization of the primary structure of a protein requires much more material than mere identification by MS sequencing of a few peptides.

To study the modifications of a single protein, chromatographic purifications, antibody precipitations, or both can be used to isolate sufficient amounts. Modern analysis methods tolerate contamination much better than earlier methods; therefore, the total amount of recovered protein is more important than absolute purity. Often, sodium dodecyl sulfate-polyacrylamide gel electrophoresis (SDS-PAGE) will be the final preparation step and researchers should attempt to isolate at least Coomassie-stainable amounts (several picomoles or 1 g) of protein to increase the chance of detecting and characterizing modifications.

7.5.1.2 PTM Mapping of a Purified Protein

Once a protein has been isolated, a variety of techniques can be used to determine the identities of modified amino acids. In some cases, the precise molecular weight of the intact protein can be established by MS, especially if the protein is not too heterogeneous, its mass is less than about 100 kDa, and it is in a buffer that is compatible with MS. Once the masses of the nonmodified and modified amino acid residues add up to the measured intact molecular weight, the protein is completely characterized.

Amino-terminal protein sequencing by the classical technique of Edman degradation is still the method of choice to determine proteolytic processing.

Carboxy-terminal processing can also be determined by amino acid sequencing, albeit at a much lower sensitivity. Detailed characterization of modification happens after enzymatic or chemical degradation of the protein. The resulting peptides are usually separated by high-performance liquid chromatography (HPLC) as described above. In Edman degradation, collected peptide fractions are applied to the sequencer and their amino acid sequence determined. Modified amino acids become apparent because of their absence or retention-time shift in the corresponding sequencing cycle. If the mass of the intact peptide has been determined, then the nature of the modification can be confidently assessed.

Often, the peptide mass pattern will hint at the nature of the modification, such as multiple mass differences of 162 Da for glycosylation or the presence of a "satellite mass" less 98 Da in the case of phospho-serine and phospho-threonine because of the elimination of phosphoric acid.

The mass of the modified peptide is usually not sufficient to determine the nature of the modification so peptides are fragmented by MS to localize the modification. In these "tandem mass spectrometry" (MS/MS) experiments, peptide ions are collided with inert gas, leading to fragmentation, usually at the peptide bonds. Some modified amino acid residues remain intact during this process. In this case, the fragmentation pattern is similar to the unmodified peptide with the difference that the location of the modified amino acid is revealed by its mass increment. Thus, ideally, the mass and location of the modification can be determined. In practice, the fragmentation pattern may or may not allow exact localization of the modification, depending on the completeness of the fragmentation pattern.

If the modification is labile, then it will be lost before the peptide itself fragments. In this case, the peptide can still be sequenced and identified, but only the mass increment (not the location of the modification) is determined. Examples of stable modifications are acetylation (+42 Da), which is found on the N termini of many proteins or on specific lysine residues, and arginine methylation (+14 Da). Examples of labile modifications are O-linked *N*-acetylglucosamine (GlcNAc; +203 Da) and sulfation (+80 Da). The phospho group (+80 Da) can be stable (e.g., in the case of phospho-tyrosine) or relatively labile (e.g., in the case of phospho-threonine and especially phospho-serine).

Modifying groups that are easily lost from the peptide can themselves be used as "reporter groups" to detect the presence of the modified peptide in several different ways: In "in-source fragmentation," excess energy in the ionization or ion-sampling process leads to the characteristic presence of the reporter ion in mass spectra. Subsequent sequencing of the peptide peaks can then identify the modified peptide. Conversely, in the "neutral loss" technique, mild collisions in the collision cell between the two sections of a tandem mass spectrometer lead to loss of the modifying group. The second mass analyzer is set at a mass offset corresponding to the mass-to-charge ratio (m/z) of the expected modification. A signal can only reach the detector if the peptides were modified and the mass changed by the expected amount during collision.

7.5.1.3 PTM Mapping of Protein Populations

Although the methods just described are very powerful for the characterization of individual, purified proteins and have helped elucidate numerous biological mechanisms, the real promise of proteomics is to assess systematically the modifications of large numbers of proteins. There are three commonly used approaches: analysis of affinity purified proteins using LC MS/MS, analysis of peptides using LC MS/MS, and derivatization-based methods.

The strategy of affinity-based enrichment of modified proteins combines established biochemical, genetic, and immunological methods for enrichment of modified-protein populations with recently developed MS techniques for protein mixture analysis. This strategy is particularly attractive because the enrichment step is often a single experiment (e.g., an immunoprecipitation) and the subsequent identification of the protein mixture is usually reduced to a single LC MS/MS experiment as well.

The phospho-proteome has been extensively explored with this strategy. For example, cells stimulated with EGF can be immunoprecipitated with anti-phosphotyrosine antibody. Another modification of great interest, the enzymatic attachment of ubiquitin to cellular proteins that marks them for destruction, is also under investigation. In an elegant experiment, yeast ubiquitin was replaced by a histidine-tagged version, allowing selective purification and identification of the ubiquitinated proteome. As these examples show, the combination of selective enrichment of modified proteins with MS mixture analysis can be very powerful. The critical step is the development of the enrichment protocol. Subsequently, the proteins have only to be identified, thus avoiding the difficulties of detailed modification mapping mentioned earlier.

Recent technological developments have made it increasingly feasible to directly analyze very complex peptide mixtures by LC MS/MS. A single chromatographic run can result in the identification of hundreds of modified peptides, especially as following metal or affinity enrichment strategies. Peptide mixtures derived from complex protein mixtures are very difficult to analyze comprehensively. If one is interested in specific modifications, the peptide complexity can be reduced by affinity methods. For example, phospho-peptides can be captured selectively through their negatively charged phospho group on immobilized-metal affinity (IMAC) columns. Recently, this technique has been made much more specific by esterifying, and thereby neutralizing, the negatively charged amino acid residues before the IMAC step, allowing identification of hundreds of phospho-peptides in yeast cell lysates. The method has also been used in combination with phospho-tyrosine protein affinity purification.

Chemical derivatization of the modifying group potentially allows attachment of a "hook" for affinity purification. For example, the phosphate group can be converted to an affinity tag by an elimination/Michael addition reaction or by phosphoamidate chemistry. It should be noted, however, that only very simple and extremely efficient chemical derivatization steps are compatible with proteomics. If any heterogeneity is introduced by the chemical reaction (e.g., as a result of <100% conversion efficiency or side reactions), the peptide samples become even more complex and it is then only possible to analyze modifications of the most abundant proteins.

7.5.2 *Identification of Protein Complexes*

Vital cellular functions such as DNA replication, transcription, and mRNA translation require the coordinated action of a large number of proteins that are assembled into an array of multiprotein complexes of distinct composition and structure. Similarly, biological processes are orchestrated and regulated by dynamic signaling networks of interacting proteins that link chemical or physical stimuli to specific effector molecules. The analysis of protein complexes and protein–protein interaction networks– and the dynamic behavior of these networks as a function of time and cell state– are therefore of central importance in biological research.

7.5.2.1 Affinity Purification

Different approaches have been used to characterize protein complexes and protein–protein interaction networks. The first interactome maps were obtained using a yeast two hybrid approach. More recently, a combination of affinity purification and mass spectrometry (AP–MS) has been used to greatly advance our understanding of protein-complex composition. With the AP–MS method, multiprotein complexes are isolated directly from cell lysates through one or more AP steps. Complex components are then identified by MS. In contrast to yeast two-hybrid and related methods, AP–MS can be performed under near physiological conditions and in the relevant organism and cell type. AP–MS does not typically perturb relevant post-translational modifications, which are often crucial for the organization and/or activity of complexes. Another advantage of AP–MS is that it can be used to probe dynamic changes in the composition of protein complexes, especially when used in combination with quantitative proteomics techniques.

Standard approaches that use affinity-tagged recombinant proteins have allowed for parallel sample preparation without the need to optimize the purification protocol for each protein complex. Proteins of interest are simply expressed in-frame with an epitope tag (at either the N or C terminus), which is then used as an affinity handle to purify the tagged protein (the bait) along with its interacting partners (the prey). Although several different tags or tag combinations have been successfully used in many low-throughput studies (see Cummings and Kornfeld 1982), high-throughput studies have primarily used either the flag or tandem affinity purification (TAP) tags.

In the flag-tag approach, C-terminally flag-tagged proteins are expressed under the control of a GAL-inducible promoter and isolated in a single step using an anti-flag antibody resin. In the TAP-tag approach genes, the proteins of interest are fused to a C-terminal dual-epitope tag via homologous recombination, such that the proteins were expressed under their own promoters. Protein purification is carried out in two steps, first via the protein A moiety in the TAP tag (which binds immunoglobulin G (IgG)–sepharose) and then via the calmodulin-binding peptide (which exhibits high affinity to calmodulin–sepharose).

The AP–MS technique generates a list of proteins detected in a given sample but does not necessarily reveal the composition of individual protein complexes. The data from a single AP–MS experiment represents an average of binding partners and protein complexes. If the bait protein is a component of multiple alternative complexes, a single AP–MS analysis cannot be used to decipher this multiplicity of associations. This is an important limitation because proteins can have dramatically different roles as components of different types of complexes. The structure of multiprotein complexes can only be revealed indirectly through high-density AP–MS approaches. However, as described below, analysis of the composition of an intact protein complex with defined biochemical properties can be used to directly reveal the composition of a given complex.

7.5.2.2 Biochemical Fractionation in Protein Complex Analysis

The fractionation approaches described above have been used widely for the separation and enrichment of protein complexes. AP of at least one of the sample components using, for example, an inhibitor or a ligand has also frequently been included in biochemical purification schemes to significantly increase enrichment. Depending on the nature of the particular protein complex, a combination of these separation methods can yield pure preparations.

Although fractionation approaches have been used successfully for the characterization of the composition of numerous biologically relevant protein assemblies, these methods are not generic and must be tailored to a particular complex of interest. This limitation prevents their application to genome-wide studies. However, combining one or more of such biochemical fractionation techniques with a generic AP protocol (such as an epitope tag) can provide a surrogate for a complete biochemical isolation of a protein complex.

Another strategy for the analysis of large multiprotein assemblies (or organelles) is to monitor co-fractionation profiles using quantitative MS and then to compare the acquired profiles with those of known components of the protein complex or organelle of interest. This can be accomplished by monitoring the number and intensity of the peptide signals for each detected protein across adjacent fractions (for example, throughout a sucrose or glycerol gradient).

7.5.2.3 Crosslinking of Protein Complexes

A problem that is encountered during the isolation of intact native protein complexes from cells or tissue is that only protein–protein interactions that are resistant to the lysis and purification conditions will survive to be detected by MS. Several different strategies have been devised to freeze transient or labile protein interactions by using chemical crosslinking reagents. Crosslinkers possess at least two reactive groups that form covalent bonds with target molecules. These reactive groups are separated by a spacer arm of a defined length (usually in the range of

5–15 Å) that determines the maximal distance between two molecules. This confers some degree of specificity to the crosslinking process: Molecules in close proximity are more likely to be crosslinked than distant species. However, protein–protein crosslinking techniques present multiple experimental and analytical challenges. The choice of crosslinker is crucial, as crosslinkers vary in cell-permeability, reactivity, and arm length. Crosslinking reaction conditions must also be closely monitored, such that bona fide protein–protein interactions are stabilized and undesired crosslinks (to contaminating proteins) are minimized.

Although many chemical crosslinkers can be used to stabilize complexes in theory, only a few have been successfully used for in vivo crosslinking followed by MS analysis. Formaldehyde and di-thiobis-succinimidyl-propionate (DSP; an amine-reactive, homobifunctional, thiol-cleavable and membrane-permeable crosslinker) have been used most often to identify novel interacting partners. Crosslinkers are also particularly attractive for revealing interactions that involve membrane proteins (or microsomes), as the detergent concentrations used to solubilize the membranes and extract the proteins typically also disrupt protein–protein interactions. Mild formaldehyde crosslinking has also been performed in animals and has allowed for the stringent immunopurification of an intact secretase complex as well as the identification of protein interactors for the cellular prion protein.

7.5.2.4 Complex Stoichiometry

Determining complex stoichiometry by mass spectrometry has thus far been challenging. However, one promising strategy to determine protein stoichiometry in a complex is to combine complex isolation with isotope-based absolute quantitative proteomics. If all of the components of a complex are known, synthetic tryptic peptides can be generated to monitor the abundance of each of the proteins in the complex. These peptides can be synthesized with heavy isotopes and then mixed with an unlabelled sample (as in the AQUA approach) or labeled in parallel to the samples (e.g., by reaction with iTRAQ).

7.6 Summary

Technologies to quantitatively interrogate proteomes are becoming a standard part of the arsenal of biologic researchers. These technologies have evolved rapidly over the past decade. Owing to the complex and dynamic nature of proteomes, it has now become clear that there is no “one-fits-all” strategy to address biological questions. While techniques for say analyzing protein complexes are already quite mature, experiments such as global protein expression profiling for biomarker discovery, are still under development. In this chapter, we attempted to provide a technical guide to the three main components of proteomics: (1) sample preparation, (2) mass spectrometric analysis, and (3) data analysis. We then highlighted two mature applications of these components to demonstrate that a fundamental canon of approaches (e.g., fractionation, followed by LCMS) can be assembled into powerful pipelines to ask and answer important biological questions.

References

Alvarez-Manilla G, Atwood Guo Y, Warren NL, Orlando R, Pierce M (2006) Tools for glycoproteomic analysis: size exclusion chromatography facilitates identification of tryptic glycopeptides with N-linked glycosylation sites. J Proteome Res 5:701–708.

Andersen JS, Wilkinson CJ, Mayor T, Mortensen P, Nigg EA, Mann M (2003) Proteomic characterization of the human centrosome by protein correlation profiling. Nature 426:570–574.

Anderson NL, Anderson NG (2002) The human plasma proteome: history, character, and diagnostic prospects. Mol Cell Proteomics 1:845–867.

Bakhtiar R, Guan Z (2005) Electron capture dissociation mass spectrometry in characterization of post-translational modifications. Biochem Biophys Res Commun 334:1–8.

Beynon RJ, Doherty MK, Pratt JM, Gaskell SJ (2005) Multiplexed absolute quantification in proteomics using artificial QCAT proteins of concatenated signature peptides. Nat Methods 2:587–589.

Blagoev B, Kratchmarova I, Ong SE, Nielsen M, Foster LJ, Mann M (2003) A proteomics strategy to elucidate functional protein–protein interactions applied to EGF signaling. Nat Biotechnol 21:315–318.

Blagoev B, Ong S-E, Kratchmarova I, Mann M (2004) Temporal analysis of phosphotyrosine-dependent signaling networks by quantitative proteomics. Nat Biotechnol 22:1139–1145.

Bondarenko PV, Chelius D, Shaler TA (2002) Identification and relative quantitation of protein mixtures by enzymatic digestion followed by capillary reversed-phase liquid chromatography–tandem mass spectrometry. Anal Chem 74:4741–4749.

Brill LM, Motamedchaboki K, Wu S, Wolf DA (2009) Comprehensive proteomic analysis of Schizosaccharomyces pombe by two-dimensional HPLC-tandem mass spectrometry. Methods 48:311–319.

Brunner E, Ahrens CH, Mohanty S, Baetschmann H, Loevenich S, Potthast F, Deutsch EW, Panse C, de Lichtenberg U, Rinner O, Lee H, Pedrioli PG, Malmstrom J, Koehler K, Schrimpf S, Krijgsveld J, Kregenow F, Heck AJ, Hafen E, Schlapbach R, Aebersold R (2007) A high-quality catalog of the *Drosophila melanogaster* proteome. Nat Biotechnol 25:576–583.

Bunkenborg J, Pilch BJ, Podtelejnikov AV, Wisniewski JR (2004) Screening for *N*-glycosylated proteins by liquid chromatography mass spectrometry. Proteomics 4:454–465.

Burke TW, Mant CT, Black JA, Hodges RS (1989) Strong cation-exchange high-performance liquid chromatography of peptides. Effect of non-specific hydrophobic interactions and linearization of peptide retention behaviour. J Chromatogr 476:377–389.

Bylund D, Danielsson R, Malmquist G, Markides KE (2002) Chromatographic alignment by warping and dynamic programming as a pre-processing tool for PARAFAC modelling of liquid chromatography–mass spectrometry data. J Chromatogr A 961:237–244.

Che F-y, Fricker LD (2002) Quantitation of neuropeptides in Cpefat/Cpefat mice using differential isotopic tags and mass spectrometry. Anal Chem 74:3190–3198.

Chelius D, Bondarenko PV (2002) Quantitative profiling of proteins in complex mixtures using liquid chromatography and mass spectrometry. J Proteome Res 1:317–323.

Chelius D, Zhang T, Wang G, Shen R-F (2003) Global protein identification and quantification technology using two-dimensional liquid chromatography nanospray mass spectrometry. Anal Chem 75:6658–6665.

Colinge J, Masselot A, Giron M, Dessingy T, Magnin J (2003) OLAV: towards high-throughput tandem mass spectrometry data identification. Proteomics 3:1454–1463.

Comisarow MB, Marshall AG (1974) Fourier transform ion cyclotron resonance spectroscopy. Chem Phys Lett 25:282–283.

Cooper HJ, Hudgins RR, Hakansson K, Marshall AG (2002) Characterization of amino acid side chain losses in electron capture dissociation. J Am Soc Mass Spectrom 13:241–249.

Craig R, Beavis RC (2004) TANDEM: matching proteins with tandem mass spectra. Bioinformatics 20:1466–1467.

Craig R, Cortens JP, Beavis RC (2005) The use of proteotypic peptide libraries for protein identification. Rapid Commun Mass Spectrom 19:1844–1850.

Craig R, Cortens JC, Fenyo D, Beavis RC (2006) Using annotated peptide mass spectrum libraries for protein identification. J Proteome Res 5:1843–1849.

Cummings RD, Kornfeld S (1982) Fractionation of asparagine-linked oligosaccharides by serial lectin-Agarose affinity chromatography. A rapid, sensitive, and specific technique. J Biol Chem 257:11235–11240.

Domon B, Aebersold R (2006) Mass spectrometry and protein analysis. Science 312:212–217.

Durr E, Yu J, Krasinska KM, Carver LA, Yates JR, Testa JE, Oh P, Schnitzer JE (2004) Direct proteomic mapping of the lung microvascular endothelial cell surface in vivo and in cell culture. Nat Biotechnol 22:985–992.

Elias JE, Gygi SP (2007) Target-decoy search strategy for increased confidence in large-scale protein identifications by mass spectrometry. Nat Methods 4:207–214.

Eng JK, McCormack AL, Yates JR (1994) An approach to correlate tandem mass spectral data of peptides with amino-acid-sequence in a protein database. J Am Soc Mass Spectrom 5:976–989.

Faca V, Pitteri SJ, Newcomb L, Glukhova V, Phanstiel D, Krasnoselsky A, Zhang Q, Struthers J, Wang H, Eng J, Fitzgibbon M, McIntosh M, Hanash S (2007) Contribution of protein fractionation to depth of analysis of the serum and plasma proteomes. J Proteome Res 6:3558–3565.

Fan X, She YM, Bagshaw RD, Callahan JW, Schachter H, Mahuran DJ (2004) A method for proteomic identification of membrane-bound proteins containing Asn-linked oligosaccharides. Anal Biochem 332:178–186.

Ficarro SB, McCleland ML, Stukenberg PT, Burke DJ, Ross MM, Shabanowitz J, Hunt DF, White FM (2002) Phosphoproteome analysis by mass spectrometry and its application to *Saccharomyces cerevisiae*. Nat Biotechnol 20:301–305.

Frank A, Pevzner P (2005) PepNovo: de novo peptide sequencing via probabilistic network modeling. Anal Chem 77:964–973.

Frewen BE, Merrihew GE, Wu CC, Noble WS, MacCoss MJ (2006) Analysis of peptide MS/MS spectra from large-scale proteomics experiments using spectrum libraries. Anal Chem 78:5678–5684.

Gao J, Opiteck GJ, Friedrichs MS, Dongre AR, Hefta SA (2003) Changes in the protein expression of yeast as a function of carbon source. J Proteome Res 2:643–649.

Geer LY, Markey SP, Kowalak JA, Wagner L, Xu M, Maynard DM, Yang X, Shi W, Bryant SH (2004) Open mass spectrometry search algorithm. J Proteome Res 3:958–964.

Gerber SA, Rush J, Stemman O, Kirschner MW, Gygi SP (2003) Absolute quantification of proteins and phosphoproteins from cell lysates by tandem MS. Proc Natl Acad Sci U S A 100:6940–6945.

Gilchrist A, Au CE, Hiding J, Bell AW, Fernandez-Rodriguez J, Lesimple S, Nagaya H, Roy L, Gosline SJ, Hallett M, Paiement J, Kearney RE, Nilsson T, Bergeron JJ (2006) Quantitative proteomics analysis of the secretory pathway. Cell 127:1265–1281.

Glocker MO, Borchers C, Fiedler W, Suckau D, Przybylski M (1994) Molecular characterization of surface topology in protein tertiary structures by amino-acylation and mass spectrometric peptide mapping. Bioconjug Chem 5:583–590.

Goodlett DR, Keller A, Watts JD, Newitt R, Yi EC, Purvine S, Eng JK, von Haller P, Aebersold R, Kolker E (2001) Differential stable isotope labeling of peptides for quantitation and de novo sequence derivation. Rapid Commun Mass Spectrom 15:1214–1221.

Goshe MB, Conrads TP, Panisko EA, Angell NH, Veenstra TD, Smith RD (2001) Phosphoprotein isotope-coded affinity tag approach for isolating and quantitating phosphopeptides in proteome-wide analyses. Anal Chem 73:2578–2586.

Goshe MB, Veenstra TD, Panisko EA, Conrads TP, Angell NH, Smith RD (2002) Phosphoprotein isotope-coded affinity tags: application to the enrichment and identification of low-abundance phosphoproteins. Anal Chem 74:607–616.

Gruhler A, Schulze WX, Matthiesen R, Mann M, Jensen ON (2005) Stable isotope labeling of *Arabidopsis thaliana* cells and quantitative proteomics by mass spectrometry. Mol Cell Proteomics 4:1697–1709.

Gygi SP, Rist B, Gerber SA, Turecek F, Gelb MH, Aebersold R (1999) Quantitative analysis of complex protein mixtures using isotope-coded affinity tags. Nat Biotechnol 17:994–999.

Hansen KC, Schmitt-Ulms G, Chalkley RJ, Hirsch J, Baldwin MA, Burlingame AL (2003) Mass spectrometric analysis of protein mixtures at low levels using cleavable C-13-isotope-coded affinity tag and multidimensional chromatography. Mol Cell Proteomics 2:299–314.

Hardman M, Makarov AA (2003) Interfacing the orbitrap mass analyzer to an electrospray ion source. Anal Chem 75:1699–1705.

Higgs RE, Knierman MD, Gelfanova V, Butler JP, Hale JE (2005) Comprehensive label-free method for the relative quantification of proteins from biological samples. J Proteome Res 4:1442–1450.

Hirabayashi J (2004) Lectin-based structural glycomics: glycoproteomics and glycan profiling. Glycoconj J 21:35–40.

Hsu J-L, Huang S-Y, Chow N-H, Chen S-H (2003) Stable-isotope dimethyl labeling for quantitative proteomics. Anal Chem 75:6843–6852.

Hsu J-L, Huang S-Y, Chen S-H (2006) Dimethyl multiplexed labeling combined with microcolumn separation and MS analysis for time course study in proteomics. Electrophoresis 27: 3652–3660.

Hu Q, Noll RJ, Li H, Makarov A, Hardman M, Graham Cooks R (2005) The Orbitrap: a new mass spectrometer. J Mass Spectrom 40:430–443.

Ishihama Y, Oda Y, Tabata T, Sato T, Nagasu T, Rappsilber J, Mann M (2005) Exponentially modified protein abundance index (emPAI) for estimation of absolute protein amount in proteomics by the number of sequenced peptides per protein. Mol Cell Proteomics 4:1265–1272.

Jaitly N, Monroe ME, Petyuk VA, Clauss TRW, Adkins JN, Smith RD (2006) Robust algorithm for alignment of liquid chromatography mass spectrometry analyses in an accurate mass and time tag data analysis pipeline. Anal Chem 78:7397–7409.

James P, Quadroni M, Carafoli E, Gonnet G (1993) Protein identification by mass profile fingerprinting. Biochem Biophys Res Commun 195:58–64.

Jensen ON, Podtelejnikov AV, Mann M (1997) Identification of the components of simple protein mixtures by high-accuracy peptide mass mapping and database searching. Anal Chem 69:4741–4750.

Ji J, Chakraborty A, Geng M, Zhang X, Amini A, Bina M, Regnier F (2000) Strategy for qualitative and quantitative analysis in proteomics based on signature peptides. J Chromatogr B Biomed Sci Appl 745:197–210.

Ji C, Guo N, Li L (2005) Differential dimethyl labeling of N-termini of peptides after guanidination for proteome analysis. J Proteome Res 4:2099–2108.

Johnson KL, Muddiman DC (2004) A method for calculating 16o/18o peptide ion ratios for the relative quantification of proteomes. J Am Soc Mass Spectrom 15:437–445.

Karas M, Hillenkamp F (1988) Laser desorption ionization of proteins with molecular masses exceeding 10,000 daltons. Anal Chem 60:2299–2301.

Keller A, Nesvizhskii AI, Kolker E, Aebersold R (2002) Empirical statistical model to estimate the accuracy of peptide identifications made by MS/MS and database search. Anal Chem 74:5383–5392.

Khalsa-Moyers G, McDonald WH (2006) Developments in mass spectrometry for the analysis of complex protein mixtures. Brief Funct Genomic Proteomic 5:98–111.

Kirkpatrick DS, Gerber SA, Gygi SP (2005) The absolute quantification strategy: a general procedure for the quantification of proteins and post-translational modifications. Methods 35: 265–273.

Kito K, Ota K, Fujita T, Ito T (2007) A synthetic protein approach toward accurate mass spectrometric quantification of component stoichiometry of multiprotein complexes. J Proteome Res 6:792–800.

Krijgsveld J, Ketting RF, Mahmoudi T, Johansen J, Artal-Sanz M, Verrijzer CP, Plasterk RHA, Heck AJR (2003) Metabolic labeling of *C. elegans* and *D. melanogaster* for quantitative proteomics. Nat Biotechnol 21:927–931.

Lam H, Deutsch EW, Eddes JS, Eng JK, King N, Stein SE, Aebersold R (2007) Development and validation of a spectral library searching method for peptide identification from MS/MS. Proteomics 7:655–667.

Li J, Steen H, Gygi SP (2003) Protein profiling with cleavable isotope-coded affinity tag (cICAT) reagents: the yeast salinity stress response. Mol Cell Proteomics 2:1198–1204.

Li X-j, Yi EC, Kemp CJ, Zhang H, Aebersold R (2005) A software suite for the generation and comparison of peptide arrays from sets of data collected by liquid chromatography–mass spectrometry. Mol Cell Proteomics 4:1328–1340.

Liu H, Sadygov RG, Yates JR (2004) A model for random sampling and estimation of relative protein abundance in shotgun proteomics. Anal Chem 76:4193–4201.

Lee YH, Han H, Chang SB, Lee SW (2004) Isotope-coded N-terminal sulfonation of peptides allows quantitative proteomic analysis with increased de novo peptide sequencing capability. Rapid Commun Mass Spectrom 18:3019–3027.

Louris JN, Cooks RG, Syka JEP, Kelley PE, Stafford GC, Todd JFJ (1987) Instrumentation, applications, and energy deposition in quadrupole ion-trap tandem mass spectrometry. Anal Chem 59:1677–1685.

Lu P, Vogel C, Wang R, Yao X, Marcotte EM (2007) Absolute protein expression profiling estimates the relative contributions of transcriptional and translational regulation. Nat Biotechnol 25:117–124.

Ma B, Zhang K, Hendrie C, Liang C, Li M, Doherty-Kirby A, Lajoie G (2003) PEAKS: powerful software for peptide de novo sequencing by tandem mass spectrometry. Rapid Commun Mass Spectrom 17:2337–2342.

Makarov A (2000) Electrostatic axially harmonic orbital trapping: a high-performance technique of mass analysis. Anal Chem 72:1156–1162.

Mallick P, Schirle M, Chen SS, Flory MR, Lee H, Martin D, Ranish J, Raught B, Schmitt R, Werner T, Kuster B, Aebersold R (2007) Computational prediction of proteotypic peptides for quantitative proteomics. Nat Biotechnol 25:125–131.

Mann M, Hojrup P, Roepstorff P (1993) Use of mass spectrometric molecular weight information to identify proteins in sequence databases. Biol Mass Spectrom 22:338–345.

March RE (1997) An introduction to quadrupole ion trap mass spectrometry. J Mass Spectrom 32:351–369.

Marko-Varga G, Lindberg H, Lofdahl CG, Jonsson P, Hansson L, Dahlback M, Lindquist E, Johansson L, Foster M, Fehniger TE (2005) Discovery of biomarker candidates within disease by protein profiling: principles and concepts. J Proteome Res 4:1200–1212.

Mason DE, Liebler DC (2003) Quantitative analysis of modified proteins by LC–MS/MS of peptides labeled with phenyl isocyanate. J Proteome Res 2:265–272.

McLafferty FW, Horn DM, Breuker K, Ge Y, Lewis MA, Cerda B, Zubarev RA, Carpenter BK (2001) Electron capture dissociation of gaseous multiply charged ions by Fourier-transform ion cyclotron resonance. J Am Soc Mass Spectrom 12:245–249.

Nesvizhskii AI, Vitek O, Aebersold R (2007) Analysis and validation of proteomic data generated by tandem mass spectrometry. Nat Methods 4:787–797.

Oda Y, Owa T, Sato T, Boucher B, Daniels S, Yamanaka H, Shinohara Y, Yokoi A, Kuromitsu J, Nagasu T (2003) Quantitative chemical proteomics for identifying candidate drug targets. Anal Chem 75:2159–2165.

Old WM, Meyer-Arendt K, Aveline-Wolf L, Pierce KG, Mendoza A, Sevinsky JR, Resing KA, Ahn NG (2005) Comparison of label-free methods for quantifying human proteins by shotgun proteomics. Mol Cell Proteomics 4:1487–1502.

Olsen JV, Andersen JR, Nielsen PA, Nielsen ML, Figeys D, Mann M, Wisniewski JR (2004) HysTag – a novel proteomic quantification tool applied to differential display analysis of membrane proteins from distinct areas of mouse brain. Mol Cell Proteomics 3:82–92.

Olsen JV, Blagoev B, Gnad F, Macek B, Kumar C, Mortensen P, Mann M (2006) Global, in vivo, and site-specific phosphorylation dynamics in signaling networks. Cell 127:635–648.

Ong SE, Blagoev B, Kratchmarova I, Kristensen DB, Steen H, Pandey A, Mann M (2002) Stable isotope labeling by amino acids in cell culture, SILAC, as a simple and accurate approach to expression proteomics. Mol Cell Proteomics 1:376–386.

Ong SE, Foster LJ, Mann M (2003) Mass spectrometric-based approaches in quantitative proteomics. Methods 29:124–130.

Ono M, Shitashige M, Honda K, Isobe T, Kuwabara H, Matsuzuki H, Hirohashi S, Yamada T (2006) Label-free quantitative proteomics using large peptide data sets generated by nanoflow liquid chromatography and mass spectrometry. Mol Cell Proteomics 5:1338–1347.

Pappin DJ, Hojrup P, Bleasby AJ (1993) Rapid identification of proteins by peptide-mass fingerprinting. Curr Biol 3:327–332.

Parish R (1989) Comparison of linear regression methods when both variables contain error: relation to clinical studies. Ann Pharmacother 23:891–898.

Park K-S, Mohapatra DP, Misonou H, Trimmer JS (2006) Graded regulation of the Kv2.1 potassium channel by variable phosphorylation. Science 313:976–979.

Perkins DN, Pappin DJ, Creasy DM, Cottrell JS (1999) Probability-based protein identification by searching sequence databases using mass spectrometry data. Electrophoresis 20:3551–3567.

Porath J (1992) Immobilized metal ion affinity chromatography. Protein Expr Purif 3:263–281.

Qian W-J, Goshe MB, Camp DG, Yu L-R, Tang K, Smith RD (2003) Phosphoprotein isotope-coded solid-phase tag approach for enrichment and quantitative analysis of phosphopeptides from complex mixtures. Anal Chem 75:5441–5450.

Ramos-Fernandez A, Lopez-Ferrer D, Vazquez J (2007) Improved method for differential expression proteomics using trypsin-catalyzed 18O labeling with a correction for labeling efficiency. Mol Cell Proteomics 6:1274–1286.

Ranish JA, Hahn S, Lu Y, Yi EC, Li XJ, Eng J, Aebersold R (2004) Identification of TFB5, a new component of general transcription and DNA repair factor IIH. Nat Genet 36:707–713.

Rao KCS, Carruth RT, Miyagi M (2005) Proteolytic 18O labeling by peptidyl-Lys metalloendopeptidase for comparative proteomics. J Proteome Res 4:507–514.

Reynolds JA, Tanford C (1970) The gross conformation of protein-sodium dodecyl sulfate complexes. J Biol Chem 245:5161–5165.

Reynolds KJ, Yao X, Fenselau C (2002) Proteolytic 18O labeling for comparative proteomics: evaluation of endoprotease Glu-C as the catalytic agent. J Proteome Res 1:27–33.

Ross PL, Huang YN, Marchese JN, Williamson B, Parker K, Hattan S, Khainovski N, Pillai S, Dey S, Daniels S, Purkayastha S, Juhasz P, Martin S, Bartlet-Jones M, He F, Jacobson A, Pappin DJ (2004) Multiplexed protein quantitation in *Saccharomyces cerevisiae* using amine-reactive isobaric tagging reagents. Mol Cell Proteomics 3:1154–1169.

Saito A, Nagasaki M, Oyama M, Kozuka-Hata H, Semba K, Sugano S, Yamamoto T, Miyano S (2007) AYUMS: an algorithm for completely automatic quantitation based on LC-MS/MS proteome data and its application to the analysis of signal transduction. BMC Bioinform 8:15.

Salomon AR, Ficarro SB, Brill LM, Brinker A, Phung QT, Ericson C, Sauer K, Brock A, Horn DM, Schultz PG, Peters EC (2003) Profiling of tyrosine phosphorylation pathways in human cells using mass spectrometry. Proc Natl Acad Sci U S A 100:443–448.

Schmidt A, Kellermann J, Lottspeich F (2005) A novel strategy for quantitative proteomics using isotope-coded protein labels. Proteomics 5:4–15.

Schnölzer M, Jedrzejewski P, Lehmann WD (1996) Protease-catalyzed incorporation of 18O into peptide fragments and its application for protein sequencing by electrospray and matrix-assisted laser desorption/ionization mass spectrometry. Electrophoresis 17:945–953.

Schuchardt S, Sickmann A (2007) Protein identification using mass spectrometry: a method overview. EXS 97:141–170.

Senko MW, Hendrickson CL, Emmett MR, Shi SD, Marshall AG (1997) External accumulation of ions for enhanced electrospray ionization fourier transform ion cyclotron resonance mass spectrometry. J Am Soc Mass Spectrom 8:970–976.

Shukla AK, Futrell JH (2000) Tandem mass spectrometry: dissociation of ions by collisional activation. J Mass Spectrom 35:1069–1090.

Sleno L, Volmer DA (2004) Ion activation methods for tandem mass spectrometry. J Mass Spectrom 39:1091–1112.

Storey JD, Tibshirani R (2003) Statistical significance for genomewide studies. Proc Natl Acad Sci U S A 100:9440–9445.

Strittmatter EF, Ferguson PL, Tang K, Smith RD (2003) Proteome analyses using accurate mass and elution time peptide tags with capillary LC time-of-flight mass spectrometry. J Am Soc Mass Spectrom 14:980–991.

Strittmatter EF, Kangas LJ, Petritis K, Mottaz HM, Anderson GA, Shen Y, Jacobs JM, Camp DG 2nd, Smith RD (2004) Application of peptide LC retention time information in a discriminant function for peptide identification by tandem mass spectrometry. J Proteome Res 3:760–769.

Syka JE, Coon JJ, Schroeder MJ, Shabanowitz J, Hunt DF (2004a) Peptide and protein sequence analysis by electron transfer dissociation mass spectrometry. Proc Natl Acad Sci U S A 101:9528–9533.

Syka JEP, Marto JA, Bai DL, Horning S, Senko MW, Schwartz JC, Ueberheide B, Garcia B, Busby S, Muratore T, Shabanowitz J, Hunt DF (2004b) Novel linear quadrupole ion trap/FT mass spectrometer: performance characterization and use in the comparative analysis of histone H3 post-translational modifications. J Proteome Res 3:621–626.

Tang H, Arnold RJ, Alves P, Xun Z, Clemmer DE, Novotny MV, Reilly JP, Radivojac P (2006) A computational approach toward label-free protein quantification using predicted peptide detectability. Bioinformatics 22:e481–e488.

Tao WA, Wollscheid B, O'Brien R, Eng JK, Li X-j, Bodenmiller B, Watts JD, Hood L, Aebersold R (2005) Quantitative phosphoproteome analysis using a dendrimer conjugation chemistry and tandem mass spectrometry. Nat Methods 2:591–598.

Thompson A, Schafer J, Kuhn K, Kienle S, Schwarz J, Schmidt G, Neumann T, Hamon C (2003) Tandem mass tags: a novel quantification strategy for comparative analysis of complex protein mixtures by MS/MS. Anal Chem 75:1895–1904.

Wada Y, Tajiri M, Yoshida S (2004) Hydrophilic affinity isolation and MALDI multiple-stage tandem mass spectrometry of glycopeptides for glycoproteomics. Anal Chem 76:6560–6565.

Wang W, Zhou H, Lin H, Roy S, Shaler TA, Hill LR, Norton S, Kumar P, Anderle M, Becker CH (2003) Quantification of proteins and metabolites by mass spectrometry without isotopic labeling or spiked standards. Anal Chem 75:4818–4826.

Wang P, Tang H, Fitzgibbon MP, Mcintosh M, Coram M, Zhang H, Yi E, Aebersold R (2007) A statistical method for chromatographic alignment of LC-MS data. Biostatistics 8:357–367.

Wolf-Yadlin A, Hautaniemi S, Lauffenburger DA, White FM (2007) Multiple reaction monitoring for robust quantitative proteomic analysis of cellular signaling networks. Proc Natl Acad Sci U S A 104:5860–5865.

Wright ME, Eng J, Sherman J, Hockenbery DM, Nelson PS, Galitski T, Aebersold R (2003) Identification of androgen-coregulated protein networks from the microsomes of human prostate cancer cells. Genome Biol 5:R4.

Wu CC, MacCoss MJ, Howell KE, Matthews DE, Yates JR (2004) Metabolic labeling of mammalian organisms with stable isotopes for quantitative proteomic analysis. Anal Chem 76:4951–4959.

Yang Z, Hancock WS (2004) Approach to the comprehensive analysis of glycoproteins isolated from human serum using a multi-lectin affinity column. J Chromatogr A 1053:79–88.

Yao X, Freas A, Ramirez J, Demirev PA, Fenselau C (2001) Proteolytic 18O labeling for comparative proteomics: model studies with two serotypes of adenovirus. Anal Chem 73:2836–2842.

Yates JR 3rd, Morgan SF, Gatlin CL, Griffin PR, Eng JK (1998) Method to compare collision-induced dissociation spectra of peptides: potential for library searching and subtractive analysis. Anal Chem 70:3557–3565.

Zhang R, Regnier FE (2002) Minimizing resolution of isotopically coded peptides in comparative proteomics. J Proteome Res 1:139–147.

Zhang R, Sioma CS, Wang S, Regnier FE (2001) Fractionation of isotopically labeled peptides in quantitative proteomics. Anal Chem 73:5142–5149.

Zhang X, Jin QK, Carr SA, Annan RS (2002) N-terminal peptide labeling strategy for incorporation of isotopic tags: a method for the determination of site-specific absolute phosphorylation stoichiometry. Rapid Commun Mass Spectrom 16:2325–2332.

Zhang H, Li X-j, Martin DB, Aebersold R (2003) Identification and quantification of N-linked glycoproteins using hydrazide chemistry, stable isotope labeling and mass spectrometry. Nat Biotechnol 21:660–666.

Zhang H, Yi EC, Li XJ, Mallick P, Kelly-Spratt KS, Masselon CD, Camp DG 2nd, Smith RD, Kemp CJ, Aebersold R (2005a) High throughput quantitative analysis of serum proteins using glycopeptide capture and liquid chromatography mass spectrometry. Mol Cell Proteomics 4:144–155.

Zhang J, Schubothe K, Li B, Russell S, Lebrilla CB (2005b) Infrared multiphoton dissociation of O-linked mucin-type oligosaccharides. Anal Chem 77:208–214.

Zubarev RA (2004) Electron-capture dissociation tandem mass spectrometry. Curr Opin Biotechnol 15:12–16.

Zubarev RA, Kelleher NL, McLafferty FW (1998) Electron capture dissociation of multiply charged protein cations. A non-ergodic process. J Am Chem Soc 120:3265–3266.

Zubarev RA, Horn DM, Fridriksson EK, Kelleher NL, Kruger NA, Lewis MA, Carpenter BK, McLafferty FW (2000) Electron capture dissociation for structural characterization of multiply charged protein cations. Anal Chem 72:563–573.

Chapter 8
Tissue Microarrays in Cancer Research

Toby C. Cornish and Angelo M. De Marzo

Abstract Tissue microarrays (TMAs) are composite tissue blocks capable of accommodating over 1,000 unique tissue cores on a single glass slide. TMAs have become widely adopted in pathology and biomarker research. This chapter briefly discusses the design and construction of TMAs, the state of TMA imaging, and current methods for the analysis and management of TMA data. A significant portion of the chapter highlights the technical challenges of using formalin-fixed, paraffin embedded tissue and analyzing tissue stained using immunohistochemistry (IHC).

8.1 Background

The tissue microarray (TMA) is the latest and most successful attempt at creating a multi-tissue block for high throughput experimentation. The multi-tissue block method was first described in 1986 by Battifora et al., who created a "sausage" block by binding irregular strips of deparaffinized tissue into a sausage casing and re-embedding the tissue in paraffin (Battifora 1986). A sausage block could contain over 100 different tissues, but identifying the disorganized tissue fragments on the slides proved difficult. In 1987, Wan et al. modified the sausage technique by introducing a hollow tube to remove cores from the donor tissue blocks (Wan et al. 1987). Up to 120 of these paraffin cores were "glued" into a bundle by gently warming them, and the parallel nature of the bundles cores made tissue identification easier. Battifora et al. revisited the technique in 1990, creating a well-organized, grid-like "checkerboard" layout by embedding layered stacks of agar slabs containing strips of tissue (Battifora and Mehta 1990). In 1998 Kononen, et al. developed the tissue microarray technique by combining the block coring method with a grid layout (Kononen et al. 1998). This novel technique produced a high-density, composite block capable of accommodating over

A.M. De Marzo (✉)
Department of Pathology, Johns Hopkins University School of Medicine, Baltimore, MD, USA
e-mail: ademarz@jhmi.edu

S. Yegnasubramanian and W.B. Isaacs (eds.), *Modern Molecular Biology*, Applied Bioinformatics and Biostatistics in Cancer Research,
DOI 10.1007/978-0-387-69745-1_8,

1,000 unique tissue cores on a single glass slide and has since been widely adopted in pathology and biomarker research.

Kononen's method for assembling TMAs has persisted virtually unchanged for over a decade, providing a convenient, high-throughput format for performing histologic experimentation on an unprecedented scale. TMAs require minimal tissue, conserve expensive or rare reagents, and eliminate the need to stain and analyze hundreds of individual slides. Most importantly perhaps, TMAs standardize experimental conditions between samples, eliminating a significant source of variation. TMAs complement other high throughput molecular techniques well and have proved particularly useful for validating candidate markers at the protein level whose expression was identified as being altered at the mRNA level using cDNA or oligonucleotide microarrays in genome wide expression profiling studies (Luo et al. 2002; Rubin et al. 2002; Simon and Sauter 2003; Watanabe et al. 2005).

This chapter briefly discusses the design and construction of TMAs, the state of TMA imaging, and current methods for the analysis and management of TMA data. A significant portion of the chapter highlights the technical challenges of using formalin-fixed, paraffin embedded tissue and analyzing tissue stained using immunohistochemistry (IHC).

8.2 Collection, Fixation, and Processing of Tissues

As we move towards the era of personalized medicine, it has become obvious that our failure to standardize tissue banking methods has introduced significant unintended molecular variability into our biospecimen repositories (Compton 2009). In 2007, the National Cancer Institute released the "NCI Best Practices for Biospecimen Resources," providing guidelines to encourage the development of adequate, standardized protocols for the collection, storage, processing and documentation of biospecimens. This effort by the NCI is intended to improve both research and clinical utility of biospecimens and is viewed as a key component for fully realizing the potential of personalized, molecular medicine.

Biospecimen variability is particularly problematic for TMA construction because the hundreds of individual tissues that compose a TMA may each have its own very "personalized" history. This is especially true for collaborative TMA projects, in which the contributing institutions may have greatly divergent protocols for handling, processing and storage of clinical specimens. A similar situation arises when clinical and research materials are comingled in the same TMA, or when a collection of donor tissues span a large period of time. For example, in one large multi-institutional TMA study of PTEN phosphatase in prostate cancer, we found that distributions of the staining intensity varied significantly by institution (Faith et al. 2005). Whether these differences relate to time of fixation differences (despite all institutions using neural buffered formalin), tissue processing differences, storage condition of paraffin blocks, and/or storage conditions of the slides is unknown. One clue that there is some relation to tissue block age came from the fact that even within our own material, there were significant differences in staining intensity between older and more recent tissue. While this is true for the antibody used in this study, the extent of the problem with IHC staining variations will obviously depend on both the given antibodies and the antigens.

Variability occurs early in the life cycle of the biospecimen. Changes in the expression pattern of some analytes may begin early in surgery, and the duration of warm ischemia influences these expression changes (Dash et al. 2002; Spruessel et al. 2004; Schlomm et al. 2008). Surgical variables are difficult to control (or even to measure) in the clinical setting, but they should be strictly managed when tissues are obtained from research animals. In the clinical setting, considerably more influence can be exerted over the collection, fixation and processing steps. Despite this, little emphasis has been placed on the suitability of the established methods for subsequent immunohistochemical and molecular analyses. Only recently did the importance of hormone receptor expression and Her-2 (Erb-b2) amplification in breast carcinoma make clinical histology labs acknowledge the need for standardized techniques.

Formalin-fixed, paraffin-embedded (FFPE) tissue remains the gold standard for histomorphology. Formalin is an aqueous solution of formaldehyde that chemically crosslinks proteins and nucleic acids. This mechanism of action raises concerns about its potential to chemically alter or mask epitopes, changing the antigenicity of tissue. The vast majority of clinical labs have standardized on 10% neutral-buffered formalin, but the timing and duration of tissue fixation can vary significantly even within the same institution. Regardless of method, fixation is essential for inactivation of proteolytic enzymes, and delays in fixation can result in an irreversible loss of epitopes. Several committees have recommended that tissue fixation should begin as soon as possible and should not be delayed more than 30–60 min after surgical removal (Werner et al. 2000; Yaziji et al. 2008). Given a formalin penetration rate of 1 mm/h and an ideal formalin to tissue ratio of at least 10:1 (v:v), large specimens present a significant challenge for rapid fixation. Therefore, it is recommended that at least a few representative, block-sized sections are placed in formalin immediately (Werner et al. 2000; Yaziji et al. 2008).

There is less of a consensus regarding the optimal duration of fixation. The formalin fixation process requires at least 24 h to reach completion, yet few specimens receive the optimal exposure to formalin (Burnett 1982; Fox et al. 1985). Recommendations for breast tissue fixation suggest from 6–48 h for Her2 testing and 8–48 h for estrogen receptor staining (Goldstein et al. 2003; Wolff et al. 2007; Yaziji et al. 2008). Excessive fixation was previously considered a major impediment to IHC staining. Yet, now that antigen retrieval techniques have matured, there seems to be little decrease in antigenicity for a number of protein markers that have been tested (e.g., ER/PR Her-2/Neu and p27) unless fixation is prolonged for perhaps many weeks (Arber 2002; De Marzo et al. 2002). Background autofluorescence may also be increased by excessive formalin fixation (Del Castillo et al. 1989). Ultimately, the sensitivity of a given epitope to fixation can only be determined empirically, and it is therefore more important to maintain a consistent protocol that ensures adequate preservation for most antigens.

After fixation, tissue undergoes stepwise dehydration in a series of alcohol-containing solutions prior to paraffin embedding. If the tissue has had inadequate formalin penetration or exposure, the dehydration process may result in partial coagulative fixation and uneven staining. Following dehydration, the tissue is clarified in xylene or a similar solvent and embedded in paraffin. Unfortunately, research on the effects of processing on IHC is lacking, so standardization and common sense are paramount in achieving consistent results. Use of uncontaminated and frequently refreshed (at least once per week) solvents is recommended, as is the avoidance of excessive heat (>56–58°C) in the embedding process (Werner et al. 2000; Yaziji et al. 2008). Newer techniques for FFPE, such as microwave processing, should be validated against traditional processing techniques before combining the tissues in a TMA.

Although it is unlikely the popularity of FFPE will diminish, alternative techniques have been proposed. Other fixatives, including ethanol, are reportedly superior to formalin for the preservation of nucleic acids and proteins (Ahram et al. 2003; Vincek et al. 2003). Coagulative fixatives also produce less autofluorescence in tissue. Ethanol fixation must be validated for each antibody, though, as we have found a number of protein biomarkers that show a marked or complete loss of immunohistochemical staining when tissues were fixed in ethanol without a cross-linking agent (A.M. De Marzo, C. Umbricht, W. Gage, J. Hicks, unpublished observations). Alternatives to paraffin include resin-embedded and frozen tissue microarrays, both of which employ solvent coagulation as a fixative (Schoenberg Fejzo and Slamon 2001; Howat et al. 2005). Although these newer techniques may be suited to prospective collection of tissues, they introduce incompatibilities with the existing wealth of FFPE archival material (Grizzle 2009).

8.3 TMA Design

The construction of TMAs is a time-consuming endeavor that benefits greatly from careful planning (Fedor and De Marzo 2005; Kajdacsy-Balla et al. 2007). TMAs are typically designed using a spreadsheet or TMA-specific software, discussed later in detail. A number of factors influence TMA design. The size of a TMA is primarily determined by the number of specimens required. In addition to specimens, space should be allotted for tissue controls, and, if desired, empty rows that separate the cores into "city blocks." When placing samples, control tissues are frequently placed in known locations, while specimens are placed randomly. TMA designs should never be symmetrical. Frequently, a small pattern of cores (a "plus" or other shape) adjacent to the "origin" corner of the array is used as a registration mark (Fig. 8.1). Finally, although the total number of cores a TMA may exceed 1,000, dividing large TMA projects into multiple TMAs may significantly reduce the construction challenges involved.

The number of cores that will fit in a standard 25 × 20 mm block depends on two factors: core diameter and core spacing. Standard punch sets are available for

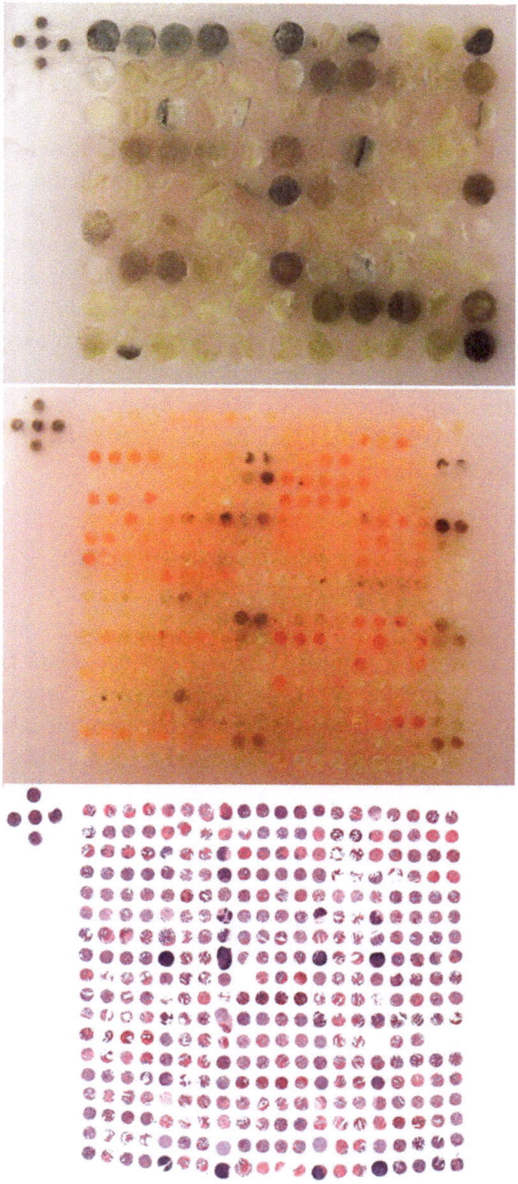

Fig. 8.1 Finished tissue microarrays. *Top*: A low density TMA block with 99 cores. Note the registration mark adjacent to the first core position. *Middle*: A higher density TMA block with 360 cores. *Bottom*: A slide cut from a high density block and stained with hematoxylin and eosin

0.6, 1.0, 1.5 and 2.0 mm diameter cores (Beecher Instruments, Sun Prairie, WI). Typical array sizes range from 400–600 cores for the 0.6 mm core to 50–100 cores for the 1.5 mm or 2.0 mm core. No rigid guidelines exist for choosing the number of replicates per sample, but the number typically ranges from 1 to 3 and should reflect both the core size and the expected heterogeneity in the tissue (Kyndi et al. 2008). Likewise, the choice of core size will be dictated by the nature of the histology being sampled. For processes that exhibit significant broad heterogeneity, three 0.6 mm cores taken from different locations in the same block may be preferable, whereas processes that have significant local variety or represent larger histologic structures (i.e., a vessel wall) are better served by fewer samples employing larger cores. The impact on the donor block can be quite significant with larger punch sizes and should also be considered, especially when such blocks are part of the clinical archive or represent rare specimens.

The spacing between cores plays an important role in the TMA density. In all cases, the spacing should be uniform across the array, and the cores should be separated by at least 0.1 mm. In practice, only the highest density TMAs require such tight spacing, and 0.2–0.4 mm is preferable for lower density arrays. Since donor cores fit tightly in the recipient block, close spacing may cause a central bulge in the recipient block, a common occurrence when constructing high density arrays. If this occurs, the block can be softened (37°C, 15 min) and gently flattened using a clean glass slide.

8.4 TMA Construction

When the design is complete and the donor blocks assembled, construction can begin on the TMA block. The TMA is constructed using punches that extract cores of paraffin-embedded tissue from the donor blocks. Specialized instruments for constructing TMAs are now available from several companies including Beecher Instruments, Veridiam (Poway, CA), Unitma (Seoul, Korea) and others. Alternatives to expensive hardware include low cost kits that use cast or prepunched recipient blocks and handheld donor punches. These kits are now available from 3DHistech (Budapest, HUNGARY), Arraymold (Cottonwood Height, Utah), Unitma, and others. There are also numerous homebrewed methods for constructing tissue microarrays (Pan et al. 2004; Pires et al. 2006; Vogel 2008). Fee-based construction services may also be available at academic centers or from private laboratories.

The manual tissue arrayer is the workhorse instrument of TMA construction (Fig. 8.2). In its simplest form, it consists of a micrometer-driven X–Y stage that translates the punch turret relative to a fixed recipient block. The punch turret carries a set of two punches designed for a particular core size. A removable bridge is placed over the recipient block to allow cores to be removed from the donor block. The general procedure for constructing a TMA with a manual arrayer is given in Box 8.1. For a detailed protocol, please refer to your manufacturer's instructions.

A number of more sophisticated instruments are now available for array construction. Additional features include integrated dissecting scopes, digital imaging, automatic positioning and computer control. Recently, the first generation of fully automated, robotic arrayers has reached the market (Beecher, 3DHistech). It should be noted that although these devices may make array construction easier, they essentially produce the exact same product.

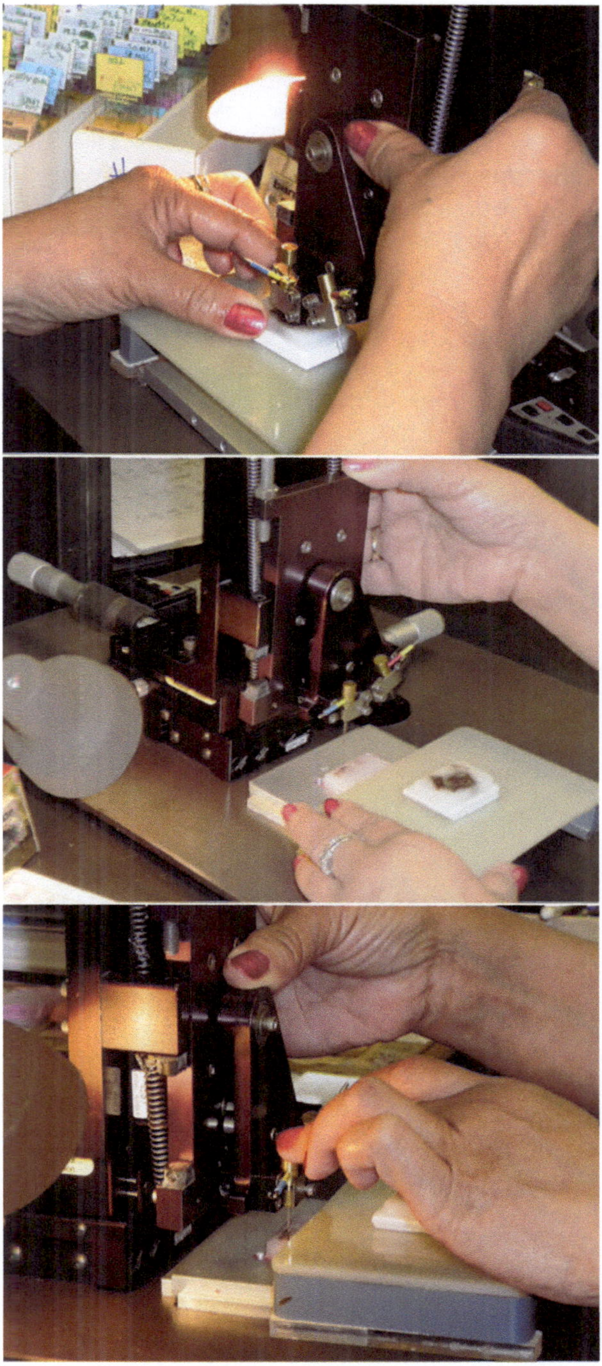

Fig. 8.2 Transferring a core using a manual arrayer. *Top*: The donor block, resting on the bridge, is manually positioned beneath the punch. Note the donor blocks and matching slides in the upper left corner. The slides and blocks are marked to indicate the area to punch. *Middle*: After the core is removed from the donor block, the bridge is retracted, revealing the recipient block, which is fixed in place. *Bottom*: The stylet is depressed, inserting the core into the recipient block (Photos courtesy Marcela "Cellie" Southerland, Johns Hopkins University, Balitmore, MD)

Box 8.1 Steps in Constructing a TMA Using a Manual Arrayer

1. Position a blank recipient block on the base plate
2. Fit a properly-sized pair of punches to the turret
3. Adjust the X–Y position until the punch is aligned with the desired location of the first core in the recipient block
4. Zero the micrometers
5. Adjust the depth stop
6. Lower the punch into the recipient block and rotate the punch slightly to free the core
7. Raise the punch and depress the stylet to discard the paraffin core
8. Rotate the turret to select the donor punch
9. Place the bridge over the recipient block
10. Manually position the donor block on the bridge, such that the area of interest is beneath the donor punch
11. Lower the punch into the donor block and rotate slightly to free the core
12. Raise the punch, then remove the bridge and donor block
13. Lower the punch and insert the donor core into the recipient block, taking care to place the core flush with the block surface
14. Advance the x-axis micrometer a distance equal to the core diameter plus the intercore space (e.g., 0.6+0.2 mm=0.8 mm)
15. Repeat for the remainder of the row
16. Return the x-axis position to zero, advance the y-axis micrometer, and begin the next row

8.5 Microtomy and Slide Storage

Once constructed, TMA blocks are precious materials. The number of slides that can be cut from a single block depends on the thickness of the tissue in the donor blocks. Typically, around 200 sections can be obtained from a single TMA block before significant core dropout occurs. The history of the block greatly influences the amount of remaining tissue, and cores from clinical tissue generally are exhausted earlier than virgin research blocks. Other factors, including the skill of the histotechnologist and the number of times the block must be faced will influence the total number of usable sections. "Facing" occurs when the block is first placed on the microtome and then continuously sectioned until the face of the block is flat, at which point complete sections can be cut. To maximize return, multiple slides are usually cut from a TMA block whenever the block is faced, or, the tape transfer method of sectioning is used. The impact of facing can be reduced by using a microtome capable of making fine adjustments to the position and angle of the block, but some amount of waste is inevitable.

Once the TMA block is faced, a quantity of slides is cut, and few, if any, sections should by discarded even if they are slightly imperfect. An extreme excess of slides should be avoided, as numerous studies have demonstrated that long term storage of FFPE tissue slides can significantly degrade the antigenicity of the tissue. Thus, the size of a batch must be balanced against potential losses during storage. Typically, around twenty sections are cut.

While there is agreement that slide storage degrades antigenicity, the rate and extent of decline varies significantly across the studies and the antibodies they tested (Jacobs et al. 1996; Bertheau et al. 1998; Wester et al. 2000). Data suggests that oxidation, heat and drying account for at least a portion of the loss (Blind et al. 2008). Storage temperature may be the single largest contributor to antigen loss. Significant losses can occur in as little as 2 weeks of storage at room temperature, while storage at 4°C appears to help only to a limited extent (van den Broek and van de Vijver 2000; Wester et al. 2000). Wester showed that storage at −20°C was the most effective method of slide storage, although there was some tissue detachment from the slides when fatty tissues were used. We have re-examined this issue very recently using stored TMA slides and found excellent antigen preservation, equal to or better than storage in the paraffin block, for 11 epitopes using non-precoated, unbaked slides stored at −20°C for up to 5 years (Berez et al. in preparation). Paraffin-coating slides, though common in practice, is of debatable value. While it should not exacerbate the problem of antigen loss, it can be problematic due to the difficulty in removing all of the paraffin from the coated slides (van den Broek and van de Vijver 2000). The combination of paraffin coating and storage in a nitrogen atmosphere is often recommended, but has only has been formally tested for up to several months (DiVito et al. 2004). In order to maximize the usable lifetime of unstained TMA slides, we currently store our TMA slides at −20°C. If older slides must be used, and loss of antigenicity is a suspected cause of false negatives, increasing antibody concentration or changing epitope retrieval methods can be attempted, but results are mixed in the literature (van den Broek and van de Vijver 2000; Wester et al. 2000; Olapade-Olaopa et al. 2001).

8.6 Commercial Slides

For simple or infrequent experiments or for those with limited access to archival tissue, commercial TMA slides are an attractive and cost-effective alternative to constructing TMA blocks. Slides are available singly or in small batches and feature related collections of normal, developmental or cancer tissues. Many commercial ventures can even provide TMAs with matched RNA or DNA. There are several potential drawbacks to commercial TMAs: the experimenter has little control over the origin, processing or quality of the tissues, and the company may provide only limited clinical and follow-up data. Typical prices for commercial TMA slides are around $200/slide for 60 cancer cases and $300/slide for 100 cases. Given the expense of TMA slides, it is best to optimize detection techniques on less expensive control tissues.

8.7 Immunodetection

In situ detection of molecules is the *raison d'être* of TMAs, and immunohistochemistry remains the primary method of detection. Other methods, including in situ hybridization (discussed briefly later) and, to a lesser extent, in situ PCR are also used, but far less frequently. Immunodetection is in fact a rather old technique, Marrack having first demonstrated in 1934 that dye-conjugated antibodies can differentially label bacteria (Marrack 1934). Coons et al. applied a refined technique to FFPE tissue, using a fluorescently labeled antibody to identify pneumococcus organisms (Coons et al. 1942). Weller and Coons then developed the indirect immunofluorescence technique, which obviated the need to label individual primary antibodies (Weller and Coons 1954). Fluorescence proved excellent for frozen tissue sections, but formalin fixation of tissue increased the autofluorescence, resulting in a poor signal-to-noise ratio. Fluorescence detection also required specialized microscopes, could not be combined with traditional H&E staining and was impermanent, fading rapidly. In 1967, Nakane and Pierce developed the now-ubiquitous horseradish peroxidase (HRP) enzyme-labeled antibody technique (Nakane and Pierce 1966). Chromogenic reporters quickly eclipsed fluorescent reagents for FFPE tissue, and the vast majority of TMA studies still employ the brown HRP substrate, 3,3′-diaminobenzidine (DAB). Other HRP substrates include the generic red substrate AEC (3-amino-9-ethylcarbazole), and the proprietary substrates NovaRed, Vector SG (blue-gray), and Vector VIP (purple) from Vector Laboratories (Burlingame, CA). Alkaline Phosphatase (AP) is the other common reporter enzyme, and AP substrates include the generic red substrate Fast Red and the proprietary substrates Vector Red, Vector Black, and Vector Blue from Vector Laboratories. Despite the expanding palette of chromogens, fluorescence detection methods, having matured significantly, are experiencing a renaissance in FFPE tissue. Compared to chromogenic IHC, immunofluorescence offers several advantages, including increased dynamic range and superior multiplexing (McCabe et al. 2005). With the recent surge in automated fluorescence slide scanners, there is a renewed debate on the relative merits of chromogenic and fluorescent detection for TMAs (Rimm 2006).

8.7.1 Immunodetection: Signal Amplification

Regardless of detection method, the efficacy of immunodetection on FFPE tissue was significantly limited until the mid-1990s. Up until that point, only 10–20% of antibodies that worked on frozen tissue sections could also be used to detect antigens on FFPE tissue sections. Synergistic advances in two methods, signal amplification and antigen retrieval, have since revolutionized immunodetection in clinical and research settings. These methods have rendered previously undetectable antigens detectable and have increased the specificity and sensitivity of others. Signal amplification and antigen retrieval are now extensively used in both chromogenic and fluorescent detection.

Signal amplification has been used with IHC for some time as even indirect methods with labeled secondary antibodies provide a small measure of signal enhancement. Greater signal amplification, however, is achieved by tethering a large complex of reporter molecules at the antigenic site. The first example of this type was the peroxidase-anti-peroxidase (PAP) procedure which allowed detection of antigens previously undetectable on FFPE tissue (Sternberger et al. 1970). PAP uses an unlabeled primary antibody, followed by a peroxidase-anti-peroxidase complex, and then a bridging antibody that links the primary and antiperoxidase antibodies. A similar method, APAAP, was developed using alkaline phosphatase as a reporter (Cordell et al. 1984).

Subsequent amplification techniques took advantage of the multivalent, high-affinity interaction of avidin and biotin. Later these techniques were adapted to use streptavidin, an uncharged protein with considerably less nonspecific binding. The first method, labeled avidin biotin (LAB), uses an unlabeled primary antibody, followed by a biotinylated secondary antibody and an HRP-conjugated avidin (Guesdon et al. 1979). The avidin–biotin complex (ABC) method was developed soon after and remains the most commonly employed amplification technique for IHC (Hsu et al. 1981). Like the LAB method, ABC uses an unlabeled primary antibody and a biotinylated secondary antibody, but this is followed by a soluble, preformed complex of avidin and biotinylated peroxidase molecules. Both the LAB and ABC methods (and their streptavidin equivalents, LSAB and SABC) are used today, and although there may be some differences in sensitivity, they are in a similar class (Giorno 1984; Sabattini et al. 1998).

A radically different approach was introduced with tyramide signal amplification (TSA), also referred to as catalyzed reporter deposition (CARD) in the literature. Commercial TSA systems are available from Perkin–Elmer/NEN Life Sciences (Boston, MA) and from Dako (Glostrup, Denmark) as the CSA System. TSA uses an unlabeled primary antibody followed by a biotinylated secondary antibody and then HRP-labeled streptavidin. Biotinylated tyramide is then added, and the HRP catalyzes its dimerization, resulting in deposition of large amounts of biotin at the site of the antigen (Bobrow et al. 1989, 1991). A reporter-labeled streptavidin is then added to the tissue for detection. The TSA method is far more sensitive than the ABC method and is believed to be the most sensitive signal amplification method currently available, but has been criticized for its complicated protocol and significant background staining. We have found, however, that background staining in TSA can be reduced significantly simply by diluting the biotinyl tyramide solution (Gurel et al. 2008).

Polymeric reporter methods are a recent development in signal amplification. They are almost as sensitive as TSA, but with "single step" amplification and less background staining. One type of polymeric reagent uses a dextran polymer backbone conjugated to numerous secondary antibodies and reporter enzymes (Heras and Roach 1995; Sabattini et al. 1998). This dextran polymer reagent is commercially available as the EnVision System from Dako. The second type of polymeric reagent is composed of directly polymerized enzymes conjugated to a secondary antibody (Shi et al. 1999; Ramos-Vara and Miller 2006). Enzyme polymer reagents are thought to suffer less steric hindrance than dextran polymer reagents and are available commercially as Powervision from Leica Biosystems (Newcastle Upon Tyne, UK) and ImmPRESS from Vector Laboratories. In addition to simplicity and high sensitivity, the polymer methods are completely biotin-free, a significant advantage for staining tissues with abundant endogenous biotin (e.g., liver, kidney and spleen). In these tissues, even pretreatment with biotin-blocking reagents cannot always eliminate nonspecific background. While both (S)ABC and L(S) AB-type methods suffer from biotin-related background, biotin-free TSA reagents are available from Dako and Perkin–Elmer.

8.7.2 Immunodetection: Antigen Retrieval

Signal amplification is an essential part of modern IHC protocols, but accessible, native epitopes are still required for the initiation of the detection scheme. Formalin fixation introduces methylene bridges between amino acid residues. This preserves structure by fixing the proteins in place, but it also alters the tertiary structure of epitopes and buries them beneath a mass of crosslinked protein (so-called "masking"). The action of formalin accounts for the significant difference in antigenicity between frozen and FFPE tissue. Antigen retrieval, the process of restoring the availability of protein epitopes, has revolutionized immunodetection in clinical and research labs (for a recent review, see D'Amico 2009). Early attempts at antigen retrieval by Huang, et al. used pronase, a proteolytic enzyme to partially digest FFPE tissue (Huang 1975). Subsequent attempts at proteolytic enzyme-induced retrieval (PIER) used pronase, proteinase K, trypsin, pepsin, and other enzymes. Although these methods were modestly successful, the use of PIER did not become widespread, and the problem of antigen retrieval persisted.

In 1991, Shi et al., revolutionized antigen retrieval by pioneering a high temperature method (Shi et al. 1997). Heat can reverse formalin crosslinking, and this is thought to be the primary mechanism in heat-induced epitope retrieval (HIER) (Fraenkel-Conrat and Olcott 1948). HIER is highly effective, yet gentle on tissue and easy to perform. For these reasons, it was quickly and widely adopted in clinical and research laboratories. Notably, HIER is less sensitive to fixation time than PIER and can retrieve antigens that were previously lost to overfixation. This has reduced many of the problems associated with variable or prolonged fixation of tissue. In addition to being effective, HIER is also inexpensive, requiring only common buffers and lab equipment. Retrieval is performed at 90–120°C, using microwave ovens, steamers, water baths, autoclaves, and pressure cookers as heat sources (Shi et al. 1997). Numerous retrieval media, ranging from heavy metal solutions to distilled water have been employed (Shi et al. 1997). Of these solutions, 10 mM citrate buffer at pH 6.0 is the most widely used and shows excellent retrieval for most antigens (Cattoretti et al. 1992). Alkaline solutions and calcium chelators may be more effective that citrate buffers for some antigens (Morgan et al. 1994; Shi et al. 1995). For these reasons, 10 mM Tris-EDTA at pH 9.0, and 10 mM Tris at pH 10 are also commonly used in HIER. Proteolytic enzymes and, less commonly, protein denaturants, such as urea, guanidine hydrochloride, guanidine thiocyanate, and formic acid are occasionally combined with HIER. In practice, the HIER protocol must be optimized for a given antibody, and common solutions like citrate buffer should be tried first.

8.7.3 Immunodetection: Validation and Controls

Immunodetection is more powerful now than it ever has been, yet mistakes in interpretation and analysis are common, producing misleading and irreproducible results. These errors can often be prevented by careful validation of antibodies and the use of proper controls. Publications that fail to report these steps should be viewed skeptically (Zha et al. 2001). The validation process for antibodies should establish the specificity of the antibody, ideally by testing it against genetically defined tissues (Gurel et al. 2008). Tissues from knock-out animals are excellent for validation, especially when knock-out animals with the reintroduced gene are also available. When knock-out animals are not available, cell lines can be easily manipulated for use in validation and as controls. For example, cells can be transfected with commercially available siRNAs for any gene of interest in order to "knock-down" its protein expression. Conversely, the gene encoding a protein of interest can be transfected into a null cell line with a known homozygous deletion, into a cell line with undetectable mRNA levels for the protein, or into a specifically targeted "knock-out" cell line, creating an excellent positive and negative pair. Cell blocks are then prepared from the cell cultures by one of several of methods. The method we often employ is to first resuspend the cells in formalin overnight at room temperature, then place them in 0.8% agarose at 42°C. When the agarose cools, it is embedded in paraffin using standard tissue processing methods (Fedor and De Marzo 2005). Depending on the cell line manipulation, a gain, loss or reduction in staining of the target should be observed. The specificity of this change can be validated by performing Western blots on the original and manipulated cell lines.

Whenever TMA slides are stained, proper controls should be included. Although these control tissues may be located on the TMA itself, they are more commonly on separate slides stained in parallel with the TMA slide. A positive control tissue (or cell block) should express the antigen of interest; a negative control tissue (or cell block) should not express the antigen of interest. As in validation, genetically defined tissues or cell lines are an excellent choice for positive and negative controls. Traditionally, a control in which the antibody is adsorbed against purified antigen is also included. Though this control can be useful, it is frequently impractical to perform and the results can be over interpreted. For example, loss of staining after adsorption only indicates that the antibody was binding to tissue using its antigen binding domain. Thus, the antibody could still be cross-reacting with another protein. This possibility can be minimized by employing the genetically defined controls discussed above and by performing a Western blot on the tissue of interest whenever possible. Finally, a control in which the primary antibody is omitted must also be performed.

8.8 TMA Imaging Systems

Although digital imaging is not essential for analyzing TMA slides, even manual scoring is greatly simplified by digitizing slides. Core images can then be directly linked to tissue, diagnosis, and scoring data. Digital imaging and image analysis software also offer the promise of fully automated TMA analysis, although this has been realized in very few cases.

Several options exist for imaging TMAs. These range from manual digital photomicrography to automated whole slide imaging (WSI). Compared to WSI, manual digital photomicrography plays a relatively minor role in bright field imaging of TMAs, but enjoys considerably more favor in fluorescence imaging, an area in which whole slide scanners have made fewer inroads. There are several advantages to manually imaging TMAs. Bright field and fluorescence microscopes are considerably more abundant than whole slide scanners and are significantly less expensive. They are also more versatile and can be outfitted with a diverse range of optics, including oil immersion, high N.A. and high power objectives – options that are less common on whole slide scanners. Because of this, high end microscopes can generally produce higher quality images than whole slide scanners. The biggest disadvantage to manually imaging TMAs is the time required to acquire the dataset. Whole slide scanners are "walk away" instruments capable of scanning a bright field slide in a few minutes and a fluorescence slide in a few hours. Manually imaging the same slide may literally take days of "hands on" time at the microscope. Slide scanners also have the advantage of being tightly coupled to TMA data management software, which adds considerable value to the process. Finally, whole slide imaging generally produces a well-calibrated, uniform digital image that is ideally suited to image analysis. If image analysis is to be performed on manually acquired digital images, care must be taken to optimize the acquisition parameters and illumination at the outset, and the images must be acquired under identical conditions for the duration of the experiment.

The availability of whole slide scanners has increased significantly, and the quality and versatility of these instruments continues to evolve with each subsequent generation. Most slide scanners are packaged with software capable of identifying TMA cores in a more or less automated way, making them ideal for digital imaging of TMAs. In general, these devices are either off-the-shelf robotic microscopes running custom software, or they are purpose-built instruments with custom hardware and software. Scanners in both categories perform well, but purpose-built instruments usually boast faster scan times and better hardware-software integration. On the other hand, scanners built around standard robotic microscopes are frequently more versatile systems.

Imaging capabilities vary greatly amongst whole slide scanners (for a review of instruments, see Rojo et al. 2006). Modified robotic microscopes that perform both bright field and fluorescence scanning are currently available from HistoRx (New Haven, Connecticut), TissueGnostics (Vienna, Austria), and Applied Imaging (San Jose, CA). Purpose-built instruments with both bright field and fluorescent capabilities are available from Hamamatsu (Hamamatsu City, Japan), BioImagene (Sunnyvale, CA), Histech3d (Budapest, Hungary) and others. Aperio (San Diego, CA), the current market leader in bright field slide scanners, has added a fluorescence-only model in ScanScope family of instruments. The cost of these instruments varies widely depending on the model, capabilities, software and hardware purchased, but an entry level configuration with a low capacity scanner, workstation, basic image database, and TMA analysis software can range from $60,000 to $250,000. Large capacity scanners with dedicated image servers and a complete image analysis toolkit capable of serving a large research group can easily cost from $300,000 to $450,000 and up. For those that wish to avoid the large capital expenditure, some WSI companies also offer scanning and analysis services, monthly leases, and other alternatives.

Finally, it is worth mentioning spectral imaging, a technique that overcomes significant problems in both bright field and fluorescence imaging. With spectral imaging, the absorbance spectra of dyes (or the emission spectra of fluorophores) are measured from reference standards, subsequently allowing mixtures of the spectra to be unmixed. Spectral imaging permits multiplexing of chromogenic dyes, improves the multiplexing of fluorescent reporters and makes true elimination of autofluoresence possible (Levenson et al. 2003; Zimmermann et al. 2003). Although promising, spectral imaging has several drawbacks. Tunable filters and spectral unmixing software are produced by very few companies, and the imaging process itself is slow and data intensive. At present, only Histech3d offers a slide scanner for automated spectral imaging of TMAs and whole slides.

8.9 Image Analysis

8.9.1 Manual Scoring

Manual scoring of TMAs is simple, inexpensive and relatively efficient. The trained human observer can easily recognize regions of interest (e.g., tissue vs matrix), diagnostic patterns (e.g., cancer vs normal), histologic features (e.g., gland vs stroma), and subcellular compartments (e.g., nucleus vs cytoplasm) where even the most sophisticated pattern recognition algorithms fail. The human visual system can also dynamically adjust to the artifacts and staining variations that inevitably occur.

There are no fixed standards for scoring of IHC, and scoring systems are generally created or adapted for the purpose at hand. Frequency-based scoring systems are amongst the simplest, counting the number of events (e.g., Ki-67 positive nuclei) in a given area or in a total number of events. Area-based scoring systems work similarly, with the observer estimating a percentage of an area that is positive. In these systems, subjectivity arises from the definition of an event, the estimation of area, and the intensity cutoff for positive staining. Although intensity information is mostly ignored, frequency and area measures produce a continuous measure of staining on a ratio scale.

In contrast, intensity-based scoring systems attempt to represent the amount of staining present. These systems frequently use a 4 or 5 category system (e.g., 0 for negative, 1 for weak, 2 for intermediate, and 3 for strong) to grade the amount of staining present in the tissue. Even with well-defined criteria, consistently assigning staining into mid-range categories can be difficult, and using more than five categories is generally counterproductive. Intensity-based scoring systems produce ordinal data that should not be treated as a continuous measure.

Hybrid scoring systems attempt to combine frequency and intensity information into a single representative score. The most prominent examples of this type are the H-score and Allred score for quantifying estrogen receptor positivity in breast cancer. The H score is calculated by multiplying the percentage of positive cells in the tissue by an intensity score (0 for negative, 1 for weak, 2 for moderate and 3 for strong), producing a score ranging from 0 to 300 (McCarty et al. 1986). The Allred score takes a different approach, summing a proportion score (0 for negative, 1 for 1/100, 2 for 1/10, 3 for 1/3, 4 for 2/3, 5 for 1/1) and an intensity score (0 for negative, 1 for weak, 2 for intermediate, and 3 for strong), producing a composite score ranging from 0 to 8 (Harvey et al. 1999). Scores of this type are ordinal data, and should not be treated as a continuous measure.

Ultimately, any manual analysis of TMAs must employ a scoring method that maximizes reproducibility and is well-suited to the research question at hand. Most importantly, the method for assigning scores needs to be well-established and formally documented prior to performing the analysis. Although there is no established requirement for multiple raters, some researchers choose to use two or three raters when performing manual analysis. When multiple observers are employed, interobserver agreement should be measured and reported. Training sets, reference slides and visual aids may significantly reduce both intra- and interobserver variability (Adams et al. 1999).

Manual scoring of IHC has been criticized for a lack of reproducibility and standardization. Although some of this criticism is valid, recently much of it has come from parties with a vested interest in instrument-based alternatives. Many of these critiques chose to ignore that pre-analytical and analytical factors may have a greater influence on reproducibility than the scoring method itself (de Jong et al. 2007). Still, valid concerns about intraobserver agreement have led to considerable reform in the guidelines for scoring ER and Her-2 scoring (Kay et al. 1994; Wolff et al. 2007; Yaziji et al. 2008). Despite criticism, and as a result of its relative ease and efficiency, manual scoring remains the gold standard for analyzing IHC.

8.9.2 *Image Analysis: Segmentation of Images*

Image analysis of TMAs is composed of two distinct steps: (1) segmentation and (2) measurement. Segmentation is the act of creating subregions in an image that represent histologically relevant categories, e.g., "cancer," "normal," "epithelium," "nucleus." While humans are particularly adept at this task, segmentation is the most challenging aspect in the automated analysis of TMAs. Although the segmentation step can be skipped entirely, including the entire TMA core may significantly degrade the fidelity of the experimental results.

Current automated segmentation techniques are computationally intense, lack precision and generalize poorly across datasets. For these reasons, manual segmentation is usually used as the first step in the analysis chain. Using simple polygon ("lasso") drawing tools, the user defines unique histologic regions of interest (ROIs). Manually defining ROIs can be labor intensive, but it is the gold standard for segmentation and is supported by almost all TMA analysis packages (Fig. 8.3).

Fully-automated segmentation of histology images is one of the few remaining obstacles to making TMAs a truly high-throughput technique. The general approaches to automatic segmentation are (1) multiplexing of stains and (2) pattern recognition.

The simplest form of stain multiplexing uses traditional counter stains, such as hematoxylin, to define a histologic ROI. Counter stains provide morphologic context for the chromogen in the foreground, and in this sense are routinely used by human observers to segment the image. Very simple forms of image segmentation can exploit this information, using, for example, the intensity of hematoxylin or DAPI to roughly define the "nuclear compartment." Automated analysis of nuclear positivity can be performed readily using this technique (Latson et al. 2003). A common oversight when using counter stains in IHC image analysis is to fail to optimize the counter stain protocol prior to staining the TMA. Suboptimal counterstaining frequently frustrates image analysis efforts.

Unfortunately, the repertoire of traditional counter stains is of limited utility as these stains define very few compartments with any specificity. To overcome this problem, antibodies that specifically label compartments of interest can be used as "counter stains." For example, an antibody against cytokeratin might be used to segment the glandular epithelium from areas of stroma. Although powerful, this technique is limited by the availability of antibodies specific for the compartment of interest. Differentiating diagnostic compartments is particularly challenging as, for example, specific anti-prostate cancer antibodies simply do not exist. Multiplexing antibodies is almost exclusively a fluorescence technique due to the challenge of unmixing spatially overlapping chromagenic stains (Rimm 2006). Quantitative unmixing of chromagens is possible with spectral imaging, though the technique is not in wide use (van der Loos 2008).

Automated pattern recognition has gained some notoriety recently from high profile applications like face recognition in crowd surveillance. Similar techniques have been applied to histologic images, and these techniques are transitioning gradually

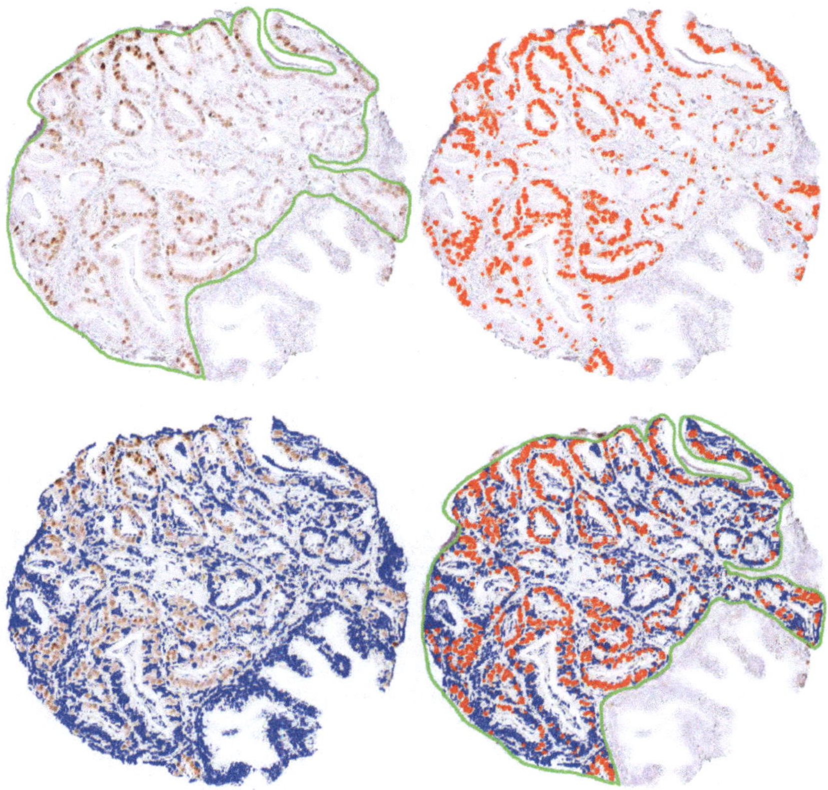

Fig. 8.3 Image analysis on a TMA core. In this example, the FrIDA/TMAJ software package has been used to measure anti-myc DAB staining with a hematoxylin counterstain. *Top left*: An ROI corresponding to the area of cancer has been manually segmented using polygon tools (green line). *Top right*: HSV color space segmentation is used to define the area of brown staining, i.e., the positive nuclei (red mask). Brown is identified by adjusting minimum and maximum values for hue, saturation and value that select pixels of the appropriate color. *Bottom left*: HSV color space segmentation is used to define the area of hematoxylin staining, i.e., the negative nuclei (blue mask). *Bottom right*: Boolean image operations are used to combine the different masks for analysis. The analysis is limited to the ROI (green line), and the percentage of positive nuclear area is calculated as Positive Nuclear Area (red mask)/Total Nuclear Area (red mask OR blue mask) (Images courtesy Dr. Bora Gurel, Johns Hopkins University, Baltimore, MD)

from research curiosities to useful segmentation tools. These tools employ a variety of image processing and statistical classification methods and have been applied to a range of research questions (Chubb et al. 2006; Doyle et al. 2006; Bilgin et al. 2007; Mete et al. 2007). The best known of these tools, GENIE, is a product of the Los Alamos National Laboratory (Los Alamos, NM) and was originally developed to classify land usage in satellite imagery (Perkins et al.2000; Brumby et al. 1999). An improved commercial version, Genie Pro, has recently been licensed by Aperio for automatic segmentation of whole slide images, including TMAs. Like other classification methods that use supervised learning, Genie requires a user-defined training set for each histologic pattern. It then uses genetic programming methods to develop an algorithm for segmenting the remaining images in the dataset. As automated tools like Genie mature, they will transform how TMAs are analyzed.

8.9.3 Measurement of Staining

Numerous commercial and noncommercial systems for measuring IHC staining are available, and describing all of them in detail is beyond the scope of this chapter (for review, see Mulrane et al. 2008). The algorithms used in these systems are usually based on a number of common techniques with occasional proprietary modifications.

Early attempts at measuring chromogenic IHC used band pass filters to separate individual stains but were frustrated by the overlapping absorbance spectra (Zhou et al. 2007). In contrast, fluorescence reporters exhibit little overlap in emission spectra, and band pass filters are used almost exclusively to separate the signals of individual signals. This simplifies measuring the staining intensity, which can then usually be done by simply measuring the grayscale intensity in the appropriate channel (Camp et al. 2002).

A number of strategies have since been employed to measure chromagenic IHC staining. The most widely applied method has been color space transformation, in which the native RGB (red-green-blue) color space of digital TMA images in transformed to one of several alternative color spaces. These have included HSV (hue-saturation-value), HSL/I/B (hue-saturation-lightness/intensity/brightness), HSD (hue-saturation-density) and Lab (luminance-color opponent a-color opponent b) (Poston and Gall 1990; Goto et al. 1992; Lamaziere et al. 1993; van Der Laak et al. 2000). Unlike RGB, these color spaces were created as analogs of human color perception, and therefore closely related colors cluster together in three-dimensional space (Russ 2007). This allows a color to be segmented by selecting a contiguous region in the color space. In this way, for example, pixels stained brown (DAB) can be separated from those that are blue (hematoxylin). Variations on color space segmentation have been used extensively in IHC analysis of tissue in both commercial and non-commercial systems (Tawfik et al. 2006; Gurel et al. 2008). Following the segmentation of stained pixels, the area of staining can be measured, the stain-positive features can be counted or the intensity of staining can be measured (Fig. 8.3). The intensity at a given pixel has been obtained using a number of methods, including the inverse grayscale value, the inverse saturation value, or an optical density (OD) calculated using Beer's law (Poston and Gall 1990; van Der Laak et al. 2000). Intensity values have also been measured directly from the yellow channel of images transformed to the CMYK (cyan-magenta-yellow-key/black) color space (Hammes et al. 2007; Pham et al. 2007).

Recently, a technique for color deconvolution from RGB images was developed and applied to the analysis of IHC. The color deconvolution algorithm uses vectors from pure stain references to create up to three independent stain channels, much like fluorescence imaging (Ruifrok and Johnston 2001). Segmentation can be performed on the appropriate stain channel, and the intensity can be measured directly from the inverse of the grayscale intensity. Color deconvolution compares favorably with HSI color space segmentation (Ruifrok et al. 2003). Both commercial and non-commercial systems for color deconvolution are available and have been used to analyze IHC on TMAs (Rabinovich et al. 2006; Halushka et al. 2010).

Despite the casual use of the term "quantitative immunohistochemistry" in the literature, very few studies actually perform quantitative measurement of the amount of antigen present in the tissue. "Semi-quantitative" is perhaps a more accurate, if somewhat nebulous, term to describe the range of analytical techniques used to measure IHC staining on TMAs (Taylor and Levenson 2006). The continuous data produced by image analysis, while more objective than manual scoring, is almost always uncalibrated data derived from measuring the products of non-linear processes, including enzymatic reactions and signal amplification steps. Uncalibrated intensity data may be measured on a continuous scale, but is not true interval data, i.e., the relative intensity units do not necessarily represent equal sized concentration intervals, making ratios of uncalibrated intensity measurements meaningless. For this reason, some researchers choose to translate the intensity data into an ordinal scoring system (e.g., 0, 1, 2, 3, 4) or to report staining in quartiles or quintiles. Attempts to incorporate standard curves for IHC are complicated by pre-analytical variables that influence staining. Approaches to calibrating IHC have included the use of tissue culture cell blocks, matrix-embedded proteins, and protein-coated beads (Riera et al. 1999; McCabe et al. 2005; Shi et al. 2005). Unfortunately, efforts to encourage calibration of IHC in the clinical and research arenas have met with little enthusiasm. In spite of these caveats, measurement of IHC remains valuable for making semi-quantitative comparisons of staining abundance.

8.10 TMA Data Management

The analysis of TMAs produces large amounts of data which can present a significant challenge to manage. A recent review of TMA data management software cataloged seventeen packages that have been described in the literature (Thallinger et al. 2007). Numerous other commercial packages are available, and each manufacturer of a TMA-capable slide scanner also offers data management software with at least limited support for TMAs (Rojo et al. 2006; Mulrane et al. 2008). In some cases, though, the software supplied with an instrument is inadequate because it lacks features that support the full lifecycle of TMA data. An ideal TMA management package should provide a secure, role-based multiuser system for storing data about donor blocks, TMA blocks, TMA sections, and analysis results. To complete the workflow, the system should also support the design of TMAs, storage of core images, and scoring of

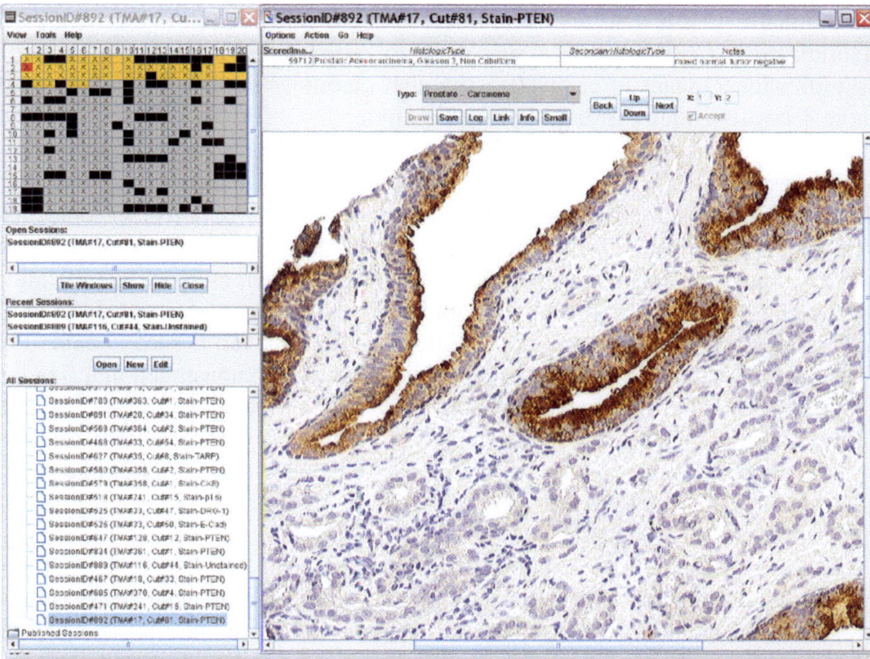

Fig. 8.4 TMA data management software. Pictured is a screenshot of TMAJ, a free and open source software package for TMA design, management and analysis. In the upper left window, a gridded view of the TMA allows for ease of navigation. The lower left window displays scoring sessions to which the user has access. The right window displays the current core (zoomed) along with histologic information and additional navigation controls

TMAs (Fig. 8.4). There are several freely available, full-featured systems designed specifically to manage TMA data. These include TMAJ (Johns Hopkins University, Baltimore, MD), and TAMEE (Graz University of Technology, Graz, Austria), both of which implement a platform-independent, client-server model and include integrated image analysis features (De Marzo et al. 2004; Thallinger et al. 2007). Although it does not integrate image analysis tools, the Stanford Tissue Microarray Database (Stanford University, Palo Alto, CA) is another full-featured system with unique data visualization tools (Marinelli et al. 2008).

8.11 DNA In Situ Hybridization

In situ hybridization (ISH) provides a means to identify genetic amplifications, deletions and rearrangements directly in tissue. Although it is less common than IHC, ISH has been used on TMAs to address a number of research questions.

Traditionally, ISH is detected using fluorescence (FISH), but chromogenic in situ hybridization (CISH) has been successful in certain applications such as Her-2 amplification (Tanner et al. 2000). Although chromogenic detection makes ISH more accessible, multiplexing of probes is limited in CISH.

ISH requires labeled DNA probes that will hybridize to complementary sequences in the FFPE tissue (for detailed protocols see Summersgill et al. 2008). These probes can directly incorporate fluorophores or incorporate other molecular tags (e.g., digoxigenin) for indirect detection. TMA slides are pretreated using proteases, and the probe is hybridized to the target. Often, a signal amplification step, using either a fluorescent or chromogenic reporter, is performed. Like IHC, ISH is believed to suffer from preanalytic variables, but fewer studies exist. For Her-2 amplification, no difference in FISH signals was demonstrated in a fixation range of 2–28 h, but signals were completely lost with prolonged (1 week) fixation (Selvarajan et al. 2002). Extended storage of blocks (>1 year) may also result in some Her-2 false negatives, while microwave processing seems to have no effect (Selvarajan et al. 2003). As with IHC, adherence to a standard fixation and processing protocol is recommended.

Analysis of ISH is similar but distinct from IHC, and can be even more time-consuming (Brown and Huntsman 2007). Some analyses require the assessment of probe colocalization (e.g., translocations), others count the number of signals (e.g., amplification), and yet others measure the intensity of signals (e.g., small sequence repeats, such as telomere length). Manual scoring of colocalization and amplification is relatively straightforward, but can be time-consuming. Manual imaging of ISH TMAs, followed by image analysis is a common method for making intensity measurements. Because ISH signals are present in multiple focal planes and require high magnification, whole slide scanners are less commonly used. Two notable exceptions are the Metafer-Metacyte system (Metasystems, Altlussheim, Germany) and the Ariol system, both of which are capable of analyzing certain types of FISH on TMAs.

8.12 Summary

The tissue microarray represents a powerful tool for high throughput analysis of tissues. It complements other high throughput techniques by providing an in situ context lacking in microarray techniques. Immunodetection and ISH techniques for TMAs are constantly evolving and cutting edge technologies like quantum dots, multispectral imaging, and automated tissue classification will enhance the impact of TMAs in the next decade. Despite coming advances, all TMA analysis techniques are subject to the same pre-analytical and analytical variables. Handling, fixation, processing and storage can have significant inadvertent effects on staining pattern and intensity long before an antibody or probe is even applied to the tissue. Likewise, inadequately controlled IHC staining experiments continue to be a major impediment to reproducibility across studies.

The TMA will not reach its full potential until biospecimen banking becomes more regimented and robust standard curves are developed to determine assay linearity and calibration.

References

Adams EJ, Green JA, Clark AH, Youngson JH (1999) Comparison of different scoring systems for immunohistochemical staining. J Clin Pathol 52:75–77.

Ahram M, Flaig MJ, Gillespie JW, Duray PH, Linehan WM, Ornstein DK, Niu S, Zhao Y, Petricoin EF 3rd, Emmert-Buck MR (2003) Evaluation of ethanol-fixed, paraffin-embedded tissues for proteomic applications. Proteomics 3:413–421.

Arber DA (2002) Effect of prolonged formalin fixation on the immunohistochemical reactivity of breast markers. Appl Immunohistochem Mol Morphol 10:183–186.

Battifora H (1986) The multitumor (sausage) tissue block: novel method for immunohistochemical antibody testing. Lab Invest 55:244–248.

Battifora H, Mehta P (1990) The checkerboard tissue block. An improved multitissue control block. Lab Invest 63:722–724.

Berez CG, Hicks JL, Lecksell K, Southerland M, Fedor H, De Marzo AM (2010) A simple storage approach for biomarker preservation in precut tissue microarray slides. Manuscript in preparation.

Bertheau P, Cazals-Hatem D, Meignin V, de Roquancourt A, Verola O, Lesourd A, Sene C, Brocheriou C, Janin A (1998) Variability of immunohistochemical reactivity on stored paraffin slides. J Clin Pathol 51:370–374.

Bilgin C, Demir C, Nagi C, Yener B (2007) Cell-graph mining for breast tissue modeling and classification. Conference proceedings. Conf Proc IEEE Eng Med Biol Soc 2007:5311–5314.

Blind C, Koepenik A, Pacyna-Gengelbach M, Fernahl G, Deutschmann N, Dietel M, Krenn V, Petersen I (2008) Antigenicity testing by immunohistochemistry after tissue oxidation. J Clin Pathol 61:79–83.

Bobrow MN, Harris TD, Shaughnessy KJ, Litt GJ (1989) Catalyzed reporter deposition, a novel method of signal amplification. Application to immunoassays. J Immunol Meth 125:279–285.

Bobrow MN, Shaughnessy KJ, Litt GJ (1991) Catalyzed reporter deposition, a novel method of signal amplification. II. Application to membrane immunoassays. J Immunol Meth 137: 103–112.

Brown LA, Huntsman D (2007) Fluorescent in situ hybridization on tissue microarrays: challenges and solutions. J Mol Histol 38:151–157.

Brumby SP, Theiler J, Perkins SJ, Harvey NR, Szymanski JJ, Bloch JJ, Mitchell M (1999) Investigation of Feature Extraction by a Genetic Algorithm. Proc SPIE. 3812:24–31.

Burnett MG (1982) The mechanism of the formaldehyde clock reaction. J Chem Educ 59:160–162.

Camp RL, Chung GG, Rimm DL (2002) Automated subcellular localization and quantification of protein expression in tissue microarrays. Nat Med 8:1323–1327.

Cattoretti G, Becker MH, Key G, Duchrow M, Schluter C, Galle J, Gerdes J (1992) Monoclonal antibodies against recombinant parts of the Ki-67 antigen (MIB 1 and MIB 3) detect proliferating cells in microwave-processed formalin-fixed paraffin sections. J Pathol 168:357–363.

Chubb C, Inagaki Y, Sheu P, Cummings B, Wasserman A, Head E, Cotman C (2006) BioVision: an application for the automated image analysis of histological sections. Neurobiol Aging 27:1462–1476.

Compton CC (2009) The surgical specimen is the personalized part of personalized cancer medicine. Ann Surg Oncol 16:2079–2080.

Coons AH, Creech HJ, Jones RN, Berliner E (1942) The Demonstration of Pneumococcal Antigen in Tissues by the Use of Fluorescent Antibody. J Immunol. 45:159–170.

Cordell JL, Falini B, Erber WN, Ghosh AK, Abdulaziz Z, MacDonald S, Pulford KA, Stein H, Mason DY (1984) Immunoenzymatic labeling of monoclonal antibodies using immune complexes of alkaline phosphatase and monoclonal anti-alkaline phosphatase (APAAP complexes). J Histochem Cytochem 32:219–229.

Dash A, Maine IP, Varambally S, Shen R, Chinnaiyan AM, Rubin MA (2002) Changes in differential gene expression because of warm ischemia time of radical prostatectomy specimens. Am J Pathol 161:1743–1748.

D'Amico F, Skarmoutsou E, Stivala F (2009) State of the art in antigen retrieval for immunohistochemistry. J Immunol Methods. Feb 28;341(1–2):1–18.

de Jong D, Rosenwald A, Chhanabhai M, Gaulard P, Klapper W, Lee A, Sander B, Thorns C, Campo E, Molina T, Norton A, Hagenbeek A, Horning S, Lister A, Raemaekers J, Gascoyne RD, Salles G, Weller E, Lunenburg Lymphoma Biomarker C (2007) Immunohistochemical prognostic markers in diffuse large B-cell lymphoma: validation of tissue microarray as a prerequisite for broad clinical applications – a study from the Lunenburg Lymphoma Biomarker Consortium. J Clin Oncol 25:805–812.

De Marzo AM, Fedor HH, Gage WR, Rubin MA (2002) Inadequate formalin fixation decreases reliability of p27 immunohistochemical staining: probing optimal fixation time using high-density tissue microarrays. Hum Pathol 33:756–760.

De Marzo AM, Morgan JD, Iacobuzio-Donahue C, Razzaque B, Faith DA (2004) TMAJ: open source software to manage a tissue microarray database. Arch Pathol Lab Med 128:1094.

Del Castillo P, Llorente AR, Stockert JC (1989) Influence of fixation, exciting light and section thickness on the primary fluorescence of samples for microfluorometric analysis. Basic Appl Histochem 33:251–257.

DiVito KA, Charette LA, Rimm DL, Camp RL (2004) Long-term preservation of antigenicity on tissue microarrays. Lab Invest 84:1071–1078.

Doyle S, Rodriguez C, Madabhushi A, Tomaszeweski J, Feldman M (2006) Detecting prostatic adenocarcinoma from digitized histology using a multi-scale hierarchical classification approach. Conf Proc IEEE Eng Med Biol Soc 1:4759–4762.

Faith DA, Ertoy-Baydar D, Spolter YS, Platz EA, Rubin MA, Ayala G, De Marzo AM (2005) Multi-institution automated image analysis of PTEN protein in prostatic adenocarcinoma. Mod Pathol 18:140A.

Fedor HL, De Marzo AM (2005) Practical methods for tissue microarray construction. Meth Mol Med 103:89–101.

Fox CH, Johnson FB, Whiting J, Roller PP (1985) Formaldehyde fixation. J Histochem Cytochem 33:845–853.

Fraenkel-Conrat H, Olcott HS (1948) Reaction of formaldehyde with proteins; cross-linking of amino groups with phenol, imidazole, or indole groups. J Biol Chem 174:827–843.

Giorno R (1984) A comparison of two immunoperoxidase staining methods based on the avidin–biotin interaction. Diagn Immunol 2:161–166.

Goldstein NS, Ferkowicz M, Odish E, Mani A, Hastah F (2003) Minimum formalin fixation time for consistent estrogen receptor immunohistochemical staining of invasive breast carcinoma. Am J Clin Pathol 120:86–92.

Goto M, Nagatomo Y, Hasui K, Yamanaka H, Murashima S, Sato E (1992) Chromaticity analysis of immunostained tumor specimens. Pathol Res Pract 188:433–437.

Grizzle W (2009) Special symposium: fixation and tissue processing models. Biotechnic and histochemistry: official publication of the Biological Stain Commission, 1–9.

Guesdon JL, Ternynck T, Avrameas S (1979) The use of avidin–biotin interaction in immunoenzymatic techniques. J Histochem Cytochem 27:1131–1139.

Gurel B, Iwata T, Koh CM, Jenkins RB, Lan F, Van Dang C, Hicks JL, Morgan J, Cornish TC, Sutcliffe S, Isaacs WB, Luo J, De Marzo AM (2008) Nuclear MYC protein overexpression is an early alteration in human prostate carcinogenesis. Mod Pathol 21:1156–1167.

Halushka MK, Cornish TC, Lu J, Selvin S, Selvin E (2010) Creation, validation, and quantitative analysis of protein expression in vascular tissue microarrays. Cardiovasc Pathol 19(3): 136–146.

Hammes LS, Korte JE, Tekmal RR, Naud P, Edelweiss MI, Valente PT, Longatto-Filho A, Kirma N, Cunha-Filho JS (2007) Computer-assisted immunohistochemical analysis of cervical cancer biomarkers using low-cost and simple software. Appl Immunohistochem Mol Morphol 15:456–462.

Harvey JM, Clark GM, Osborne CK, Allred DC (1999) Estrogen receptor status by immunohistochemistry is superior to the ligand-binding assay for predicting response to adjuvant endocrine therapy in breast cancer. J Clin Oncol 17:1474–1481.

Heras A, Roach CM (1995) Enhanced polymer detection system for immunohistochemistry. Mod Pathol 8:165A.

Howat WJ, Warford A, Mitchell JN, Clarke KF, Conquer JS, McCafferty J (2005) Resin tissue microarrays: a universal format for immunohistochemistry. J Histochem Cytochem 53: 1189–1197.

Hsu SM, Raine L, Fanger H (1981) Use of avidin–biotin–peroxidase complex (ABC) in immunoperoxidase techniques: a comparison between ABC and unlabeled antibody (PAP) procedures. J Histochem Cytochem 29:577–580.

Huang SN (1975) Immunohistochemical demonstration of hepatitis B core and surface antigens in paraffin sections. Lab Invest 33:88–95.

Jacobs TW, Prioleau JE, Stillman IE, Schnitt SJ (1996) Loss of tumor marker-immunostaining intensity on stored paraffin slides of breast cancer. J Natl Cancer Inst 88:1054–1059.

Kajdacsy-Balla A, Geynisman JM, Macias V, Setty S, Nanaji NM, Berman JJ, Dobbin K, Melamed J, Kong X, Bosland M, Orenstein J, Bayerl J, Becich MJ, Dhir R, Datta MW, Cooperative Prostate Cancer Tissue R (2007) Practical aspects of planning, building, and interpreting tissue microarrays: the cooperative prostate cancer tissue resource experience. J Mol Histol 38:113–121.

Kay EW, Walsh CJ, Cassidy M, Curran B, Leader M (1994) C-erbB-2 immunostaining: problems with interpretation. J Clin Pathol 47:816–822.

Kononen J, Bubendorf L, Kallioniemi A, Barlund M, Schraml P, Leighton S, Torhorst J, Mihatsch MJ, Sauter G, Kallioniemi OP (1998) Tissue microarrays for high-throughput molecular profiling of tumor specimens. Nat Med 4:844–847.

Kyndi M, Sorensen FB, Knudsen H, Overgaard M, Nielsen HM, Andersen J, Overgaard J (2008) Tissue microarrays compared with whole sections and biochemical analyses.. A subgroup analysis of DBCG 82 b&c. Acta Oncol (Stockholm, Sweden 47:591–599.

Lamaziere JM, Lavallee J, Zunino C, Larrue J (1993) Semiquantitative study of the distribution of two cellular antigens by computer-directed color analysis. Lab Invest 68:248–252.

Latson L, Sebek B, Powell KA, Latson L, Sebek B, Powell KA (2003) Automated cell nuclear segmentation in color images of hematoxylin and eosin-stained breast biopsy. Anal Quant Cytol Histol 25:321–331.

Levenson RM, Bearman GH, Mahadevan-Jansen A (2003) Spectral imaging: instrumentation, applications, and analysis II. Proc SPIE 4959:27–33.

Luo J, Zha S, Gage WR, Dunn TA, Hicks JL, Bennett CJ, Ewing CM, Platz EA, Ferdinandusse S, Wanders RJ, Trent JM, Isaacs WB, De Marzo AM (2002) Alpha-methylacyl-CoA racemase: a new molecular marker for prostate cancer. Cancer Res 62:2220–2226.

Marinelli RJ, Montgomery K, Liu CL, Shah NH, Prapong W, Nitzberg M, Zachariah ZK, Sherlock GJ, Natkunam Y, West RB, van de Rijn M, Brown PO, Ball CA (2008) The Stanford tissue microarray database. Nucleic Acids Res 36:D871–D877.

Marrack J (1934) The nature of antibodies. Nature 133:292–293.

McCabe A, Dolled-Filhart M, Camp RL, Rimm DL (2005) Automated quantitative analysis (AQUA) of in situ protein expression, antibody concentration, and prognosis. J Natl Cancer Inst 97:1808–1815.

McCarty KS Jr, Szabo E, Flowers JL, Cox EB, Leight GS, Miller L, Konrath J, Soper JT, Budwit DA, Creasman WT (1986) Use of a monoclonal anti-estrogen receptor antibody in the immunohistochemical evaluation of human tumors. Cancer Res 46:4244s–4248s.

Mete M, Xu X, Fan CY, Shafirstein G (2007) Automatic delineation of malignancy in histopathological head and neck slides. BMC Bioinformatics 8(Suppl 7):S17.

Morgan JM, Navabi H, Schmid KW, Jasani B (1994) Possible role of tissue-bound calcium ions in citrate-mediated high-temperature antigen retrieval. J Pathol 174:301–307.

Mulrane L, Rexhepaj E, Penney S, Callanan JJ, Gallagher WM (2008) Automated image analysis in histopathology: a valuable tool in medical diagnostics. Expert Rev Mol Diagn 8:707–725.

Nakane PK, Pierce GB Jr (1966) Enzyme-labeled antibodies: preparation and application for the localization of antigens. J Histochem Cytochem 14:929–931.

National Cancer Institute Best Practices for Biospecimen Resources (2007) US Department of Health and Human Services, Bethesda, MD.

Olapade-Olaopa EO, Ogunbiyi JO, MacKay EH, Muronda CA, Alonge TO, Danso AP, Moscatello DK, Sandhu DP, Shittu OB, Terry TR, Wong AJ, Habib FK (2001) Further characterization of storage-related alterations in immunoreactivity of archival tissue sections and its implications for collaborative multicenter immunohistochemical studies. Appl Immunohistochem Mol Morphol 9:261–266.

Pan CC, Chen PC, Chiang H (2004) An easy method for manual construction of high-density tissue arrays. Appl Immunohistochem Mol Morphol 12:370–372.

Perkins S, Theiler J, Brumby SP, Harvey NR, Porter RB, Szymanski JJ, Bloch JJ (2000) GENIE – A Hybrid Genetic Algorithm for Feature Classification in Multi-Spectral Images. Proc SPIE. 4120:52–62.

Pham NA, Morrison A, Schwock J, Aviel-Ronen S, Iakovlev V, Tsao MS, Ho J, Hedley DW (2007) Quantitative image analysis of immunohistochemical stains using a CMYK color model. Diagnos Pathol 2:8.

Pires AR, Andreiuolo Fda M, de Souza SR (2006) TMA for all: a new method for the construction of tissue microarrays without recipient paraffin block using custom-built needles. Diagnost Pathol 1:14.

Poston RN, Gall NP (1990) Hue-saturation-intensity color image analysis for the quantitation of immunoperoxidase staining. Acto Histochem Cytochem 23:730.

Rabinovich A, Krajewski S, Krajewska M, Shabaik A, Hewitt SM, Belongie S, Reed JC, Price JH (2006) Framework for parsing, visualizing and scoring tissue microarray images. IEEE Trans Inf Technol Biomed 10:209–219.

Ramos-Vara JA, Miller MA (2006) Comparison of two polymer-based immunohistochemical detection systems: ENVISION+ and ImmPRESS. J Microsc 224:135–139.

Riera J, Simpson JF, Tamayo R, Battifora H (1999) Use of cultured cells as a control for quantitative immunocytochemical analysis of estrogen receptor in breast cancer. The Quicgel method. Am J Clin Pathol 111:329–335.

Rimm DL (2006) What brown cannot do for you. Nat Biotechnol 24:914–916.

Rojo MG, Garcia GB, Mateos CP, Garcia JG, Vicente MC (2006) Critical comparison of 31 commercially available digital slide systems in pathology. Int J Surg Pathol 14:285–305.

Rubin MA, Zhou M, Dhanasekaran SM, Varambally S, Barrette TR, Sanda MG, Pienta KJ, Ghosh D, Chinnaiyan AM (2002) Alpha-Methylacyl coenzyme A racemase as a tissue biomarker for prostate cancer. JAMA: J Am Med Assoc 287:1662–1670.

Ruifrok AC, Johnston DA (2001) Quantification of histochemical staining by color deconvolution. Anal Quant Cytol Histol 23:291–299.

Ruifrok AC, Katz RL, Johnston DA (2003) Comparison of quantification of histochemical staining by hue-saturation-intensity (HSI) transformation and color-deconvolution. Appl Immunohistochem Mol Morphol 11:85–91.

Russ JC (2007) The image processing handbook. CRC Press/Taylor & Francis, Boca Raton.

Sabattini E, Bisgaard K, Ascani S, Poggi S, Piccioli M, Ceccarelli C, Pieri F, Fraternali-Orcioni G, Pileri SA (1998) The EnVision++ system: a new immunohistochemical method for diagnostics and research. Critical comparison with the APAAP, ChemMate, CSA, LABC, and SABC techniques. J Clin Pathol 51:506–511.

Schlomm T, Nakel E, Lubke A, Buness A, Chun FK, Steuber T, Graefen M, Simon R, Sauter G, Poustka A, Huland H, Erbersdobler A, Sultmann H, Hellwinkel OJ (2008) Marked gene transcript level alterations occur early during radical prostatectomy. Eur Urol 53:333–344.

Schoenberg Fejzo M, Slamon DJ (2001) Frozen tumor tissue microarray technology for analysis of tumor RNA, DNA, and proteins. Am J Pathol 159:1645–1650.

Selvarajan S, Bay BH, Choo A, Chuah KL, Sivaswaren CR, Tien SL, Wong CY, Tan PH (2002) Effect of fixation period on HER2/neu gene amplification detected by fluorescence in situ hybridization in invasive breast carcinoma. J Histochem Cytochem 50: 1693–1696.

Selvarajan S, Bay BH, Mamat SB, Choo A, Chuah KL, Sivaswaren CR, Tien SL, Wong CY, Tan PH (2003) Detection of HER2/neu gene amplification in archival paraffin-embedded breast cancer tissues by fluorescence in situ hybridization. Histochem Cell Biol 120:251–255.

Shi SR, Imam SA, Young L, Cote RJ, Taylor CR (1995) Antigen retrieval immunohistochemistry under the influence of pH using monoclonal antibodies. J Histochem Cytochem 43:193–201.

Shi SR, Cote RJ, Taylor CR (1997) Antigen retrieval immunohistochemistry: past, present, and future. J Histochem Cytochem 45:327–343.

Shi SR, Guo J, Cote RJ, Young LL, Hawes D, Shi Y, Thu S, Taylor CR (1999) Sensitivity and detection efficiency of a novel two-step detection system (power vision) for immunohistochemistry. Appl Immunohistochem Mol Morphol 7:201–208.

Shi SR, Liu C, Perez J, Taylor CR (2005) Protein-embedding technique: a potential approach to standardization of immunohistochemistry for formalin-fixed, paraffin-embedded tissue sections. J Histochem Cytochem 53:1167–1170.

Simon R, Sauter G (2003) Tissue microarray (TMA) applications: implications for molecular medicine. Expert Rev Mol Med 5:1–12.

Spruessel A, Steimann G, Jung M, Lee SA, Carr T, Fentz AK, Spangenberg J, Zornig C, Juhl HH, David KA (2004) Tissue ischemia time affects gene and protein expression patterns within minutes following surgical tumor excision. Biotechniques 36:1030–1037.

Sternberger LA, Hardy PH Jr, Cuculis JJ, Meyer HG (1970) The unlabeled antibody enzyme method of immunohistochemistry: preparation and properties of soluble antigen–antibody complex (horseradish peroxidase-antihorseradish peroxidase) and its use in identification of spirochetes. J Histochem Cytochem 18:315–333.

Summersgill B, Clark J, Shipley J (2008) Fluorescence and chromogenic in situ hybridization to detect genetic aberrations in formalin-fixed paraffin embedded material, including tissue microarrays. Nat Protoc 3:220–234.

Tanner M, Gancberg D, Di Leo A, Larsimont D, Rouas G, Piccart MJ, Isola J (2000) Chromogenic in situ hybridization: a practical alternative for fluorescence in situ hybridization to detect HER-2/neu oncogene amplification in archival breast cancer samples. Am J Pathol 157: 1467–1472.

Tawfik OW, Kimler BF, Davis M, Donahue JK, Persons DL, Fan F, Hagemeister S, Thomas P, Connor C, Jewell W, Fabian CJ (2006) Comparison of immunohistochemistry by automated cellular imaging system (ACIS) versus fluorescence in-situ hybridization in the evaluation of HER-2/neu expression in primary breast carcinoma. Histopathology 48:258–267.

Taylor CR, Levenson RM (2006) Quantification of immunohistochemistry – issues concerning methods, utility and semiquantitative assessment II. Histopathology 49:411–424.

Thallinger GG, Baumgartner K, Pirklbauer M, Uray M, Pauritsch E, Mehes G, Buck CR, Zatloukal K, Trajanoski Z (2007) TAMEE: data management and analysis for tissue microarrays. BMC Bioinform 8:81.

van den Broek LJ, van de Vijver MJ (2000) Assessment of problems in diagnostic and research immunohistochemistry associated with epitope instability in stored paraffin sections. Appl Immunohistochem Mol Morphol 8:316–321.

van Der Laak JA, Pahlplatz MM, Hanselaar AG, de Wilde PC (2000) Hue-saturation-density (HSD) model for stain recognition in digital images from transmitted light microscopy. Cytometry 39:275–284.

van der Loos CM (2008) Multiple immunoenzyme staining: methods and visualizations for the observation with spectral imaging. J Histochem Cytochem 56:313–328.

Vincek V, Nassiri M, Nadji M, Morales AR (2003) A tissue fixative that protects macromolecules (DNA, RNA, and protein) and histomorphology in clinical samples. Lab Invest 83:1427–1435.

Vogel UF (2008) Simple, inexpensive and precise paraffin tissue microarrays constructed with predrilled ordinary steel embedding moulds. Histopathology 52:255–256.

Wan WH, Fortuna MB, Furmanski P (1987) A rapid and efficient method for testing immunohistochemical reactivity of monoclonal antibodies against multiple tissue samples simultaneously. J Immunol Meth 103:121–129.

Watanabe A, Cornelison R, Hostetter G (2005) Tissue microarrays: applications in genomic research. Expert Rev Mol Diagn 5:171–181.

Weller TH, Coons AH (1954) Fluorescent antibody studies with agents of varicella and herpes zoster propagated in vitro. Proc Soc Exp Biol Med 86:789–794.

Werner M, Chott A, Fabiano A, Battifora H (2000) Effect of formalin tissue fixation and processing on immunohistochemistry. Am J Surg Pathol 24:1016–1019.

Wester K, Wahlund E, Sundstrom C, Ranefall P, Bengtsson E, Russell PJ, Ow KT, Malmstrom PU, Busch C (2000) Paraffin section storage and immunohistochemistry. Effects of time, temperature, fixation, and retrieval protocol with emphasis on p53 protein and MIB1 antigen. Appl Immunohistochem Mol Morphol 8:61–70.

Wolff AC, Hammond ME, Schwartz JN, Hagerty KL, Allred DC, Cote RJ, Dowsett M, Fitzgibbons PL, Hanna WM, Langer A, McShane LM, Paik S, Pegram MD, Perez EA, Press MF, Rhodes A, Sturgeon C, Taube SE, Tubbs R, Vance GH, van de Vijver M, Wheeler TM, Hayes DF, American Society of Clinical Oncology/College of American Pathology (2007) American Society of Clinical Oncology/College of American Pathologists guideline recommendations for human epidermal growth factor receptor 2 testing in breast cancer. Arch Pathol Lab Med 131:18–43.

Yaziji H, Taylor CR, Goldstein NS, Dabbs DJ, Hammond EH, Hewlett B, Floyd AD, Barry TS, Martin AW, Badve S, Baehner F, Cartun RW, Eisen RN, Swanson PE, Hewitt SM, Vyberg M, Hicks DG, Members of the Standardization Ad-Hoc Consensus C (2008) Consensus recommendations on estrogen receptor testing in breast cancer by immunohistochemistry. Appl Immunohistochem Mol Morphol 16:513–520.

Zha S, Gage WR, Sauvageot J, Saria EA, Putzi MJ, Ewing CM, Faith DA, Nelson WG, De Marzo AM, Isaacs WB (2001) Cyclooxygenase-2 is up-regulated in proliferative inflammatory atrophy of the prostate, but not in prostate carcinoma. Cancer Res 61:8617–8623.

Zhou L, Hodeib M, Abad JD, Mendoza L, Kore AR, Hu Z (2007) New tissue microarray technology for analyses of gene expression in frozen pathological samples. Biotechniques 43: 101–105.

Zimmermann T, Rietdorf J, Pepperkok R (2003) Spectral imaging and its applications in live cell microscopy. FEBS Lett 546:87–92.

Index

MIX
Papier aus verantwortungsvollen Quellen
Paper from responsible sources
FSC® C105338

If you have any concerns about our products,
you can contact us on
ProductSafety@springernature.com

In case Publisher is established outside the EU,
the EU authorized representative is:
Springer Nature Customer Service Center GmbH
Europaplatz 3, 69115 Heidelberg, Germany

Printed by Libri Plureos GmbH
in Hamburg, Germany